2

DEFENSE, SECURITY AND STRATEGIES

DEPARTMENT OF HOMELAND SECURITY MANAGEMENT OF AMMUNITION AND FIREARMS

SELECT ANALYSES

DEFENSE, SECURITY AND STRATEGIES

Additional books in this series can be found on Nova's website
under the Series tab.

Additional e-books in this series can be found on Nova's website
under the e-book tab.

DEFENSE, SECURITY AND STRATEGIES

DEPARTMENT OF HOMELAND SECURITY MANAGEMENT OF AMMUNITION AND FIREARMS

SELECT ANALYSES

LEWIS CALVIN
EDITOR

NOVINKA

New York

For permission to use material from this book please contact us:
Telephone 631-231-7269; Fax 631-231-8175
Web Site: http://www.novapublishers.com

Library of Congress Cataloging-in-Publication Data

ISBN: 978-1-63321-162-9

Published by Nova Science Publishers, Inc. † New York

CONTENTS

PREFACE

The Department of Homeland Security's (DHS) mission includes law enforcement elements, which require the department, through its components, to manage various types of firearms. DHS's annual ammunition purchases have declined since fiscal year 2009 and are comparable in number to the Department of Justice's (DOJ) ammunition purchases. In fiscal year 2013, DHS purchased 84 million rounds of ammunition, which is less than DHS's ammunition purchases over the past 5 fiscal years. DHS component officials said the decline in ammunition purchases in fiscal year 2013 was primarily a result of budget constraints, which meant reducing the number of training classes, and drawing on their ammunition inventories. This book addresses trends in DHS's ammunition purchases since fiscal year 2008, how DHS's purchases compare with DOJ's, and what factors affect DHS's purchase decisions.

Chapter 1 – DHS and its components have homeland security and law enforcement missions that require agents and officers to carry and be proficient in the use of firearms. DHS has more than 70,000 firearm-carrying personnel—the most of any department. DOJ has the next largest with approximately 69,000 firearm-carrying personnel. GAO was asked to examine DHS's ammunition purchases and management of ammunition and firearms. This report addresses trends in DHS's ammunition purchases since fiscal year 2008, how DHS's purchases compare with DOJ's, and what factors affect DHS's purchase decisions. GAO analyzed data from six DHS and three DOJ components that have law enforcement missions, require agents and officers to carry firearms, and purchase ammunition themselves or through their respective departments. Specifically, GAO analyzed data on ammunition purchases, usage, costs, and inventories, among other things, for fiscal years 2008 through 2013 for DHS, and for fiscal years 2011 through 2013 for DOJ.

GAO assessed the reliability of these data and found them sufficiently reliable. Data on DOJ ammunition purchases prior to fiscal year 2011 were not readily available; therefore, GAO excluded them, as discussed in the report.

Chapter 2 – This memorandum addresses recent ammunition solicitations issued and contracts awarded by components of the Department of Homeland Security (DHS). This memorandum provides information to clarify the size and scope of select acquisitions that have garnered public attention, and provides information to place those solicitations and contracts in context. This memorandum is not intended as an exhaustive survey of non-military acquisition of ammunition by the federal government, nor is it an authoritative assessment of the ammunition requirements of any particular component of the federal government.

Chapter 3 – The Department of Homeland Security Office of Inspector General (OIG) was established by the *Homeland Security Act of 2002* (Public Law 107-296) by amendment to the *Inspector General Act of 1978*. This is one of a series of audit, inspection, and special reports prepared as part of our oversight responsibilities to promote economy, efficiency, and effectiveness within the department. This report addresses the efficacy of the Department of Homeland Security's management and oversight of its components to ensure that personnel are sufficiently safeguarding and controlling firearms. It is based on interviews with employees and officials of relevant agencies and institutions, direct observations, and a review of applicable documents. The recommendations herein have been developed to the best knowledge available to our office, and have been discussed in draft with those responsible for implementation. We trust this report will result in more effective, efficient, and economical operations. We express our appreciation to all of those who contributed to the preparation of this report.

In: Department of Homeland Security … ISBN: 978-1-63321-162-9
Editor: Lewis Calvin

Chapter 1

DEPARTMENT OF HOMELAND SECURITY: AMMUNITION PURCHASES HAVE DECLINED SINCE 2009*

United States Government Accountability Office

WHY GAO DID THIS STUDY

DHS and its components have homeland security and law enforcement missions that require agents and officers to carry and be proficient in the use of firearms. DHS has more than 70,000 firearm-carrying personnel—the most of any department. DOJ has the next largest with approximately 69,000 firearm-carrying personnel. GAO was asked to examine DHS's ammunition purchases and management of ammunition and firearms.

This report addresses trends in DHS's ammunition purchases since fiscal year 2008, how DHS's purchases compare with DOJ's, and what factors affect DHS's purchase decisions. GAO analyzed data from six DHS and three DOJ components that have law enforcement missions, require agents and officers to carry firearms, and purchase ammunition themselves or through their respective departments. Specifically, GAO analyzed data on ammunition purchases, usage, costs, and inventories, among other things, for fiscal years 2008 through 2013 for DHS, and for fiscal years 2011 through 2013 for DOJ.

* This is an edited, reformatted and augmented version of United States Government Accountability Office Publication, No. GAO-14-119, dated January 2014.

GAO assessed the reliability of these data and found them sufficiently reliable. Data on DOJ ammunition purchases prior to fiscal year 2011 were not readily available; therefore, GAO excluded them, as discussed in the report.

What GAO Recommends

GAO is not making any recommendations.

What GAO Found

The Department of Homeland Security's (DHS) annual ammunition purchases have declined since fiscal year 2009 and are comparable in number to the Department of Justice's (DOJ) ammunition purchases. In fiscal year 2013, DHS purchased 84 million rounds of ammunition, which is less than DHS's ammunition purchases over the past 5 fiscal years, as shown in the figure below. DHS component officials said the decline in ammunition purchases in fiscal year 2013 was primarily a result of budget constraints, which meant reducing the number of training classes, and drawing on their ammunition inventories. From fiscal years 2008 through 2013, DHS purchased an average of 109 million rounds of ammunition for training, qualification, and operational needs, according to DHS data. DHS's ammunition purchases over the 6-year period equates to an average of 1,200 rounds purchased per firearm-carrying agent or officer per year. Over the past 3 fiscal years (2011-2013), DHS purchased an average of 1,000 rounds per firearm-carrying agent or officer and selected DOJ components purchased 1,300 rounds per firearm-carrying agent or officer.

DHS ammunition purchases are driven primarily by firearm training and qualification requirements. Most DHS firearm-carrying personnel are required to qualify four times per year, though requirements vary by component, as do the number of rounds of ammunition typically used for training and qualification. DHS components also reported considering other factors when making ammunition purchase decisions, such as changes in firearms, usage rates, and ammunition inventories. DHS components maintain inventories of ammunition to help ensure they have sufficient ammunition for the training and operational needs of their officers, as there can be months-long delays between placing an order for ammunition and receiving it. As of October

2013, DHS estimates it had approximately 159 million rounds in inventory, enough to last about 22 months to meet the training and operational needs of its firearm-carrying personnel. Ammunition inventory data provided by DOJ components indicated that inventory ranged from about 13 months' worth to about 20 months' worth.

ABBREVIATIONS

ATF	Bureau of Alcohol, Tobacco, Firearms and Explosives
CBP	U.S. Customs and Border Protection
DEA	Drug Enforcement Administration
DHS	Department of Homeland Security
DOJ	Department of Justice
FACTS	Firearms, Armor, and Credentials Tracking System
FAMS	Federal Air Marshal Service
FAR	Federal Acquisition Regulations
FBI	Federal Bureau of Investigation
FEMA	Federal Emergency Management Agency
FFDO	federal flight deck officer
FLETC	Federal Law Enforcement Training Centers
FPDS	Federal Procurement Data System
FPS	Federal Protective Service
HSAR	Homeland Security Acquisition Regulation
HSAM	Homeland Security Acquisition Manual
ICE	U.S. Immigration and Customs Enforcement
IDIQ	indefinite delivery, indefinite quantity mm millimeter
NFTTU	National Firearms Tactical and Training Unit
NPPD	National Protection and Programs Directorate
TSA	Transportation Security Administration
USSS	U.S. Secret Service
WACC	Weapons and Ammunition Commodity Council

January 13, 2014

The Honorable Tom A. Coburn, M.D.
Ranking Member
Committee on Homeland Security and
Governmental Affairs United States Senate

The Honorable Michael T. McCaul
Chairman
Committee on Homeland Security
House of Representatives

The Honorable Jeff Duncan
Chairman
Subcommittee on Oversight and Management Efficiency
Committee on Homeland Security
House of Representatives

The Honorable Richard Hudson
Chairman
Subcommittee on Transportation Security
Committee on Homeland Security
House of Representatives

The Honorable John L. Mica
Member
House of Representatives

The Department of Homeland Security (DHS) and five of its components—U.S. Customs and Border Protection (CBP), National Protection and Programs Directorate/Federal Protective Service (FPS), U.S. Immigration and Customs Enforcement (ICE), Transportation Security Administration (TSA), and the U.S. Secret Service (USSS)— have homeland security and law enforcement responsibilities and missions that require law enforcement agents and officers to carry and be proficient in the use of firearms and ammunition.[1] DHS employs more firearm-carrying law enforcement personnel than any other department— more than 70,000, according to DHS data, and DHS components purchase ammunition for these personnel.[2] Firearm-carrying personnel at DHS include Border Patrol agents, federal air marshals, and Secret Service agents, among others. In addition, another DHS component—the four interagency training sites that constitute the Federal Law Enforcement Training Centers (FLETC)—also purchases ammunition. FLETC trains an average of over 66,000 federal, state, local, tribal, and international law enforcement personnel every year, according to DHS officials. Ammunition purchased by FLETC is utilized by all students who train on firearms at its facilities. For comparison, the Department of Justice (DOJ) and its components—including the Bureau of Alcohol, Tobacco, Firearms and Explosives (ATF); Federal Bureau of Investigation (FBI); U.S. Marshals Service; Drug Enforcement Administration (DEA); and the Federal Bureau of Prisons—have the second largest number of federal personnel authorized to carry firearms, with approximately 69,000 agents and officers authorized to carry firearms as of October 2013, according to DOJ data.

At both DHS and DOJ, federal law enforcement agents and officers are authorized to carry firearms, and are required to train in their use and pass certain firearms qualification standards. Firearms proficiency is a key component of an officer's ability to successfully fulfill the component's mission and protect life and public safety. To help ensure the firearms proficiency of their firearm-carrying workforce, these agencies purchase a variety of types of ammunition for training, qualifications, and operational purposes.

Because DHS's law enforcement missions require the department, through its components, to purchase ammunition and manage various types of firearms, effective management and oversight of ammunition and firearms is an important aspect of DHS's overall mission. It is incumbent upon agencies to ensure the effective and efficient management of ammunition and firearms as stewards of taxpayer resources. You asked us to examine DHS's ammunition purchases and its management of both ammunition and firearms. This report addresses the following questions:

1) What are the trends in DHS's ammunition purchases since fiscal year 2008, what factors affect its purchase decisions, and how do DHS's purchases compare with DOJ's?
2) What policies and guidance does DHS have for managing firearms and ammunition inventories?

To address our first question, we obtained available data from DHS and DOJ law enforcement components with firearm-carrying personnel regarding their ammunition purchases, costs, usage, inventory, and the size of the authorized firearm-carrying workforce. We also obtained data on DHS planned ammunition purchases for fiscal year 2014. Specifically, we selected all DHS law enforcement components with firearm-carrying personnel—CBP, ICE, TSA, USSS, FPS, as well as FLETC.[3] We excluded the U.S. Coast Guard—a DHS component with firearm-carrying personnel—because the U.S. Coast Guard does not procure ammunition through DHS; rather, the U.S. Coast Guard procures its ammunition through Department of Defense contracts. Similarly, we selected all DOJ law enforcement components with firearm-carrying personnel. However, the Federal Bureau of Prisons could not provide ammunition purchase data comparable to that of the other components based on its method of record keeping and data retention policy. We determined that data from the Drug Enforcement Administration (DEA) were not sufficiently reliable for our purposes, because DEA could only provide estimates of purchases. Therefore, the DOJ components included in our review are ATF, FBI, and U.S. Marshals Service. We included these DOJ law enforcement agencies in our review to provide context to help understand DHS's ammunition purchases relative to those of other federal law enforcement agencies.

We selected fiscal year 2008 as the starting point for our analysis because data on DHS ammunition purchases, cost, and usage were not readily available or reliable prior to 2008, according to DHS officials. Nevertheless, obtaining data going back 5 fiscal years provides historical perspective on DHS's ammunition purchases and use over time, and allows for comparisons with purchases in fiscal year 2013 and planned purchases for fiscal year 2014. We analyzed the data to identify any changes in the amount of ammunition purchased or used year to year and interviewed knowledgeable agency officials to understand the reasons why changes occurred. For each fiscal year, we calculated the amount of ammunition purchased per agent and officer authorized to carry firearms at each DHS and DOJ component in our review, using component-provided data on the amount of ammunition purchased and the number of firearm-carrying personnel. We calculated the average number of rounds purchased per agent or officer per year for fiscal years 2008 through 2013 at DHS, excluding FLETC, by determining the average amount of ammunition purchased across DHS law enforcement components in each fiscal year and dividing that by the average number of agents or officers at DHS for each fiscal year. We determined that ammunition purchase data for DOJ components were not sufficiently reliable prior to 2011 for all components, so we used data from fiscal years 2011 through 2013 to do the same calculation. We analyzed the amount of ammunition purchased per agent or officer authorized to carry firearms to account for the varying sizes of the different departments and components. We selected ammunition purchases by DOJ to provide a comparison with DHS's ammunition purchases as DOJ has the second largest number of personnel authorized to carry firearms. However, our DOJ analysis is not to suggest that the department represents a model or standard against which DHS is assessed. Differences in the amount of ammunition purchased per firearm-carrying agent or officer reflect a number of factors, including unique mission requirements and training needs.

To assess the reliability of these data, we reviewed DHS ammunition procurement contract files and interviewed knowledgeable officials at DHS and DOJ responsible for collecting and reporting these data. In addition, for the data we obtained from DHS components on ammunition purchases and costs, we compared the data with corroborating evidence, such as information from the Federal Procurement Data System (FPDS), to determine data consistency and reasonableness. We found the data on ammunition purchase, usage, cost, and inventory, as well as workforce data, to be sufficiently reliable for our purposes, with caveats and exceptions noted throughout the report. We also interviewed DHS component officials from each of the components in our review responsible for overseeing the ammunition procurement process and determining ammunition requirements at the component level. We also discussed the purpose of quality testing the ammunition, and to understand the context of why proficiency among firearm-carrying personnel is important. Finally, for additional context, we visited the ICE National Firearms Testing Lab, which performs quality control testing on ammunition and firearms, and maintains an inventory of ammunition and firearms for distribution to DHS field locations.

To address our second question, we reviewed DHS-wide and DHS component-level policy documents, such as acquisition and personal property management directives, and firearms and ammunition guidance and operating procedures. We interviewed knowledgeable officials from the DHS Office of the Chief Procurement Officer, CBP, FLETC, FPS, ICE, TSA, and USSS regarding their management and oversight of ammunition and firearms and the related policies and procedures for DHS and its components. We also reviewed available DHS Inspector General audit and component inspection reports related to the oversight over ammunition and firearms. We did not assess if the components' policies, procedures, and process for managing their firearms and ammunition inventories were working as intended, as that was outside the scope of our review. For more information on our scope and methodology, see appendix I.

We conducted this performance audit from May 2013 to January 2014 in accordance with generally accepted government auditing standards. Those standards require that we plan and perform the audit to obtain sufficient, appropriate evidence to provide a reasonable basis for our findings and conclusions based on our audit objectives. We believe that the evidence obtained provides a reasonable basis for our findings and conclusions based on our audit objectives.

BACKGROUND

DHS and DOJ Components in Our Review with Firearm-Carrying Personnel

DHS and DOJ have several components with law enforcement functions whose personnel are authorized to carry firearms in support of accomplishing their respective missions. Table 1 describes the various law enforcement and homeland security missions of the DHS and DOJ components within our review, as well as the number of personnel authorized to carry firearms in fiscal year 2013.

Table 1. Law Enforcement Missions of Selected Department of Homeland Security (DHS) and Department of Justice (DOJ) Components and Number of Personnel Authorized to Carry Firearms in Fiscal Year 2013

Department and component	Law enforcement mission	Number ofpersonnelauthorized tocarry firearms infiscal year 2013
DHS		
U.S. Customs and Border Protection (CBP)	CBP's priority mission is to keep terrorists and their weapons out of the United States, while upholding a responsibility for securing and facilitating trade and travel. CBP's law enforcement functions include enforcing a wide range of U.S. regulations, including immigration and customs laws.	43,000a
Federal Law Enforcement Training Centers (FLETC)	FLETC provides career-long training to law enforcement professionals to help them fulfill their responsibilities safely and proficiently. FLETC has four training sites that train an average of about 66,000 federal, state, local, tribal, and international law enforcement personnel every year.	16

Department and component	Law enforcement mission	Number ofpersonnelauthorized tocarry firearms infiscal year 2013
Federal Protective Service (FPS)	A subcomponent of the National Protection and Programs Directorate, FPS is a federal law enforcement agency that provides integrated security and law enforcement services to the General Service Administration's federally owned and leased buildings, facilities, properties, and other federal properties.	955
U.S. Immigration and Customs Enforcement (ICE)	ICE is the principal investigative arm of DHS and promotes homeland security and public safety through the enforcement of federal criminal and civil laws governing border control, customs, trade, and immigration.	13,234
Transportation Security Administration (TSA)	TSA protects the nation's transportation systems to ensure freedom of movement for people and commerce. The largest firearm-carrying components of TSA are the Federal Air Marshals Service (FAMS) and federal flight deck officers (FFDOs). Federal air marshals deploy on U.S. aircraft world-wide, and conduct protection, response, detection, and assessment activities in airports and other transportation systems. Under the federal flight deck officer program, eligible flight crew members are authorized by TSA to use firearms to defend the flight deck against an act of criminal violence or air piracy.	The number ofFAMS andFFDOs isconsideredsensitive securityinformation
U.S. Secret Service (USSS)	USSS safeguards the nation's financial infrastructure and payment systems to preserve the integrity of the economy and protects national leaders; visiting heads of state; as well as government, designated sites, and national special security events.	5,014

Table 1. (Continued)

Department and component	Law enforcement mission	Number ofpersonnelauthorized tocarry firearms infiscal year 2013
DOJ		
Federal Bureau of Investigation (FBI)	The FBI protects and defends the United States against terrorist and foreign intelligence threats; upholds and enforces the criminal laws of the United States; and provides leadership and criminal justice services to federal, state, municipal, and international agencies and partners.	13,856
Bureau of Alcohol, Tobacco, Firearms, and Explosives (ATF)	ATF protects communities from violent criminals, criminal organizations, the illegal use and trafficking of firearms, the illegal use and storage of explosives, acts of arson and bombings, acts of terrorism, and the illegal diversion of alcohol and tobacco products.	3,051
U.S. Marshals Service	The U.S. Marshals Service protects the federal judiciary, apprehends federal fugitives, manages and sells seized assets acquired by criminals through illegal activities, houses and transports federal prisoners, and operates the Witness Security Program.	9,235

Source: GAO analysis of DHS and DOJ information.

[a]The data from CBP include a rounded number of personnel for one of CBP's sub-components. Therefore, we report CBP's data rounded to the thousands.

In support of their law enforcement missions, both DHS and DOJ law enforcement officers and agents use a number of different types of firearms, which require a variety of ammunition. Examples of firearms include side arms, such as pistols, and long guns, such as rifles. Commonly used ammunition includes .40 caliber, .223 caliber, and 9 millimeter, according to our analysis of data provided by DHS. More examples of commonly used firearms and ammunition can be found in appendix II.[4]

DHS Procurement of Ammunition

Example of an ICE Strategic Sourcing Contract

ICE has a strategic sourcing contract for .40 caliber ammunition that was initiated in January 2012 and will expire in January 2017. This is an indefinite delivery, indefinite quantity contract with base year and options for 4 additional years, from which other components, such as TSA and CBP, can place purchase orders to fulfill their .40 caliber ammunition needs. The maximum number of rounds of ammunition on the contract, if all 5 years are optioned, is 450 million rounds for a total cost of $109 million.

Source: GAO analysis of DHS data.

In 2003, DHS began its strategic sourcing program to leverage its buying power and secure competitive prices for a variety of goods and services, resulting in cost savings through collective procurement actions (e.g., buying in bulk quantities at lower prices).[5] DHS developed its first strategically sourced ammunition procurement in 2005. DHS's strategic sourcing contract vehicles include contracts or agreements that have been established for use by two or more components, and these types of contracts have been used by DHS components to procure ammunition. The Office of Management and Budget's Office of Federal Procurement Policy has cited DHS's efforts among best practices for implementing federal strategic sourcing initiatives, and we have also reported on DHS's strategic sourcing initiatives.[6]

DHS procures ammunition using two types of contracts—strategic sourcing and individual contracts. The decision of whether to use a strategic sourcing contract is driven by collective component needs, according to DHS officials. That is, if more than one component needs a specific type of ammunition, then that ammunition procurement is a candidate for strategic sourcing. Most of DHS's ammunition contracts, whether strategically sourced or individual contracts, are indefinite delivery, indefinite quantity (IDIQ) contracts, which are typically negotiated for a base year with additional options for purchasing ammunition up to a certain maximum number of rounds, or contract ceiling. These IDIQ contracts allow components to lock in the price, specifications, delivery costs, and other requirements and then place purchase orders throughout the negotiated time frame of the contract (e.g., indefinite delivery) for varying quantities as needed (e.g., indefinite quantity), rather than placing a single order for large amounts of ammunition. DHS orders based on the contracts as needed and pays for the ammunition on delivery. DHS is required to buy only a minimum that represents about 1 month of the projected DHS requirement, according to DHS officials.

In August 2012, DHS began requiring components to use strategic sourcing contract vehicles for procurements, including ammunition, unless procurements met certain exceptions, such as specialized types of ammunition not commonly used across components or ammunition requiring certain technical specifications.[7] Ammunition that is either not commonly used or has specifications needed by a single component can be acquired through individual component contracts according to DHS officials. Within the DHS Office of the Chief Procurement Officer, the Strategic Sourcing Program Office helps components develop, implement, and maintain sourcing strategies to enhance acquisition efficiency. According to DHS officials, the strategic sourcing process for procuring ammunition for multiple components has saved an estimated $2 million since fiscal year 2008.[8] See appendix III for a complete list of all active DHS ammunition contracts, as of October 2013.

Given the number of law enforcement and security personnel across the department, DHS established the Weapons and Ammunition Commodity Council (WACC) in October 2003 to consolidate weapons, ammunition, and other enforcement equipment requirements. While DHS components are responsible for determining their own ammunition requirements and needs, WACC is to serve as the coordinating mechanism for DHS components' ammunition and weapons procurements. Rather than have each component procure ammunition and weapons individually or draw exclusively on its own history for best practices, the members of WACC, as shown in figure 1, are to meet monthly to explore ways to maximize procurement savings, according to DHS officials. According to DHS officials, to further coordinate ammunition and weapons procurements, DHS is exploring the feasibility of broadening the scope of WACC to also track and approve ammunition purchases across the department. As of October 2013, DHS said these plans are in the development stage.

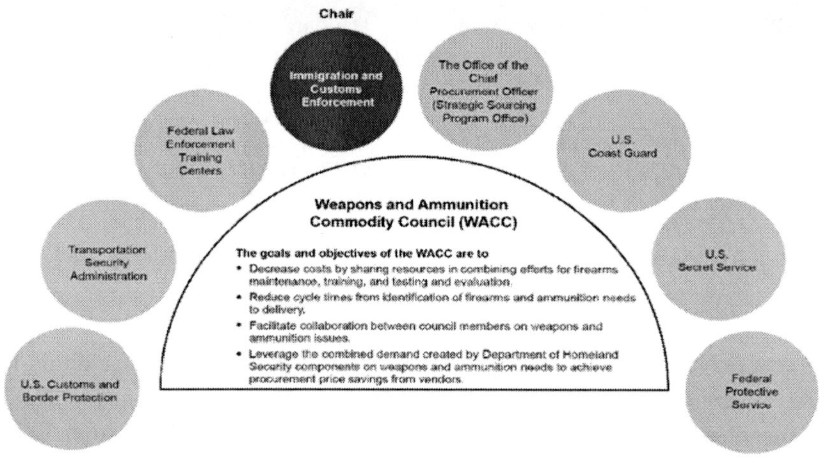

Source: DHS.

Figure 1. Members and Goals of the Department of Homeland Security's (DHS) Weapons and Ammunition Commodity Council.

DHS Organizational Responsibilities for Firearm and Ammunition Inventory Management

DHS's Office of the Chief Readiness Support Officer, within the Management Directorate, is responsible for department-wide asset management. Within each component, property management officers or firearm program managers are responsible for the component's firearm program, including ammunition. In general, component firearm instructors or firearms and weapons custodians are responsible for maintaining oversight for the shipment, receipt, issuance and periodic inventory of firearms and ammunition. At the headquarters level, the DHS Assistant Deputy for Mobile Assets and Personal Property has overall responsibility for the oversight and management of DHS firearm assets. This includes ensuring that components have an approved firearm asset management system of record and documented firearm accountability policies and procedures. The Chief Readiness Support Officer disseminates DHS's asset management program requirements, provides oversight of the program, and sets department policy. The DHS Office of the Chief Readiness Support Officer's Sensitive Asset Manager's responsibilities include evaluating, auditing, and assessing component-level firearm asset management systems of record and accountability programs to ensure compliance with laws, regulations, policies, and directives, and working with components to develop standard and uniform DHS-wide firearms accountability policies and guidelines.

DHS's Ammunition Purchases Have Declined in Recent Years and Vary Based on Numerous Factors; Annual Ammunition Purchases are Comparable to DOJ's

In fiscal year 2013, DHS purchased 84 million rounds of ammunition for its authorized firearm-carrying workforce, which is less than DHS's ammunition purchases each year over the past 5 fiscal years. DHS ammunition purchases are driven primarily by the firearms training and qualification requirements for the firearm-carrying workforce, though other factors are also considered by DHS when making ammunition purchase decisions.[9] For selected DOJ components, for fiscal years 2011 through 2013, the average number of rounds of ammunition purchased per authorized firearm-carrying personnel per year was comparable to that for DHS law enforcement components.[10] All DOJ components in our review identified similar considerations in determining their annual ammunition requirements.

DHS's Ammunition Purchases Have Decreased since Fiscal Year 2009 and Vary by Component

From fiscal year 2008 through fiscal year 2013, DHS purchased an average of 109 million rounds of ammunition per year. Yearly purchases ranged from a high of 133 million rounds in fiscal year 2009 to a low of 84 million rounds in fiscal year 2013. In comparison, the total consumer ammunition market for 2012 was approximately 9.5 billion rounds, according to the National Shooting Sports Foundation.[11] DHS's ammunition purchases over the 6-year period equate to an average of 1,200 rounds of ammunition purchased per agent or officer per year.[12] The annual total cost for these ammunition purchases ranged from $19 million to $34 million per year, with an average of $29 million for the 6- year time period.

In fiscal year 2013, DHS purchased 84 million rounds of ammunition for its authorized firearm-carrying workforce, equaling an average of 900 rounds per agent or officer that year, and a total annual cost of $19 million. According to senior DHS officials, the decline in ammunition purchases in fiscal year 2013 is the result of budget constraints, reduced training, and in one case the expiration of an ammunition contract. These officials said that although ammunition purchases declined in fiscal year 2013, DHS components relied on their ammunition inventories to maintain basic qualification and operational needs. In fiscal year 2014, DHS plans to purchase about 75 million rounds of ammunition (see fig. 2). According to DHS contract data as of October 1, 2013, the 29 existing DHS ammunition contracts extend over the next 4 fiscal years and have a remaining contract limit of approximately 704 million rounds (for various ammunition types), if every option for purchasing ammunition were exercised into fiscal year 2018. The total contract dollar ceiling on these 29 active contracts is about $285 million. The approximately 704 million rounds of ammunition represent the limit on the combined active contracts, against which orders from the manufacturers may be placed over the next several fiscal years. See appendix III for a complete list of all active DHS ammunition contracts, as of October 1, 2013.

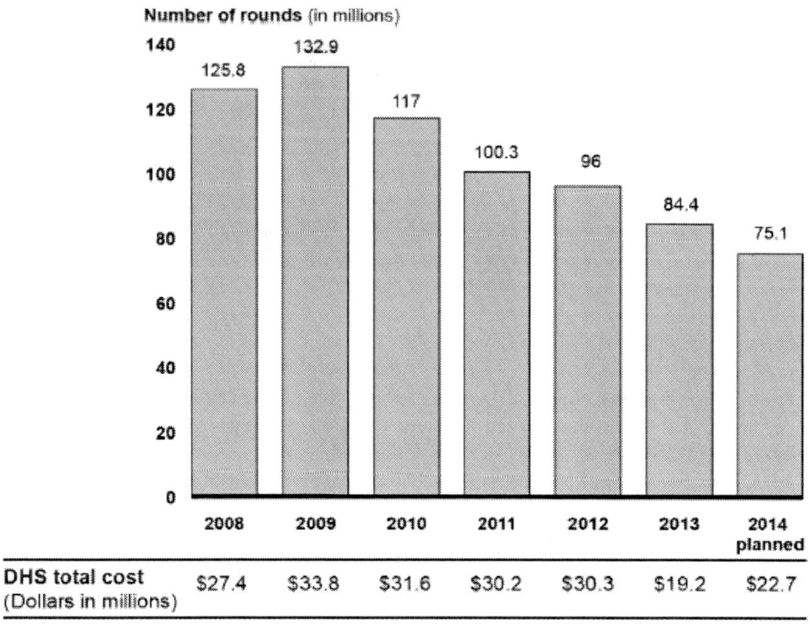

Source: GAO analysis of DHS data.

Figure 2. Total Department of Homeland Security (DHS) Ammunition Purchases for Components in Our Review from Fiscal Years 2008 through 2013.

Although the amounts of DHS's overall ammunition purchases have fallen since fiscal year 2009, the yearly changes in the amount of ammunition purchased by its component agencies have varied, as shown in figure 3. CBP—which has the most firearm-carrying personnel of the DHS components—purchased the largest amount of ammunition, on average, from fiscal year 2008 through 2013, accounting for approximately 46 percent of DHS's average purchases of ammunition for this time period. The average amount of ammunition purchased by components from fiscal years 2008 through 2013 ranged from approximately 50 million rounds of ammunition for CBP to a low of approximately 1.9 million rounds of ammunition for FPS.[13]

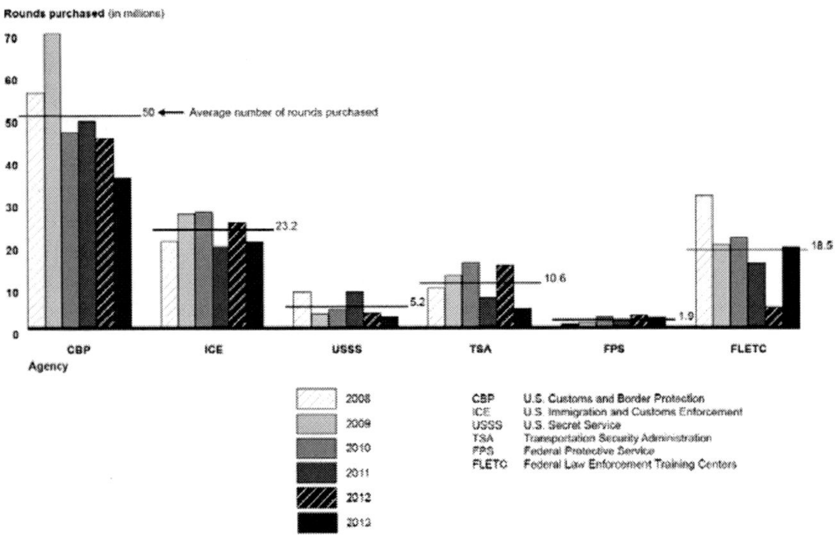

Source: GAO analysis of DHS data.

Figure 3. Number of Rounds of Ammunition Purchased, by Department of Homeland Security (DHS) Component in Our Review, from Fiscal Years 2008 through 2013.

A number of factors contribute to variation in ammunition purchases from year to year by component, such as changes in the size of the firearm-carrying workforce. For example, CBP and ICE increased their firearm-carrying workforces in fiscal years 2008 and 2009, which also increased ammunition needs for those components, as new hires needed to be trained to gain firearms proficiency prior to entering the field for duty. According to CBP data, new Border Patrol agents each use approximately 3,300 rounds during training and qualification, compared with experienced officers who might use about 600 rounds. DHS components provided data on the amount of ammunition typically used by a new law enforcement hire, which ranged from 2,000 to 5,000 rounds. From fiscal years 2008 through 2013, the DHS components in our review trained thousands of new law enforcement agents and officers, according to DHS data. Other factors that account for changes in ammunition purchases year to year include qualification and training requirements and amount of ammunition in inventory, as discussed later in this report. In addition, it is important to note that components purchase ammunition throughout the year, and orders placed in 1 fiscal year might not arrive or be used until future years, which can also contribute to variation in purchases from year to year.

Table 2. Annual Average Number of Rounds of Ammunition Purchased, per Firearm-Carrying Agent or Officer, by Department of Homeland Security (DHS) Law Enforcement Component in Our Review, Fiscal Years 2008-2013

Department of Homeland Security component	Annual average number of rounds of ammunitionpurchased per agent or officer, fiscal years 2008-2013
U.S. Customs and Border Protection	1,200
Immigration and Customs Enforcement	1,700
U.S. Secret Service	1,000
Transportation Security Administration (TSA)	[a]
Federal Protective Service (FPS)	2,000 [b]

Source: GAO analysis based on DHS data.

Notes:Table 2 does not include rounds of ammunition purchased by Federal Law Enforcement Training Centers (FLETC) for training of law enforcement officers. While DHS law enforcement officers take training at FLETC, the numbers in table 2 reflect only those rounds of ammunition purchased by components for their officers.

[a]The annual average number of ammunition rounds purchased per TSA firearm-carrying officer is considered sensitive security information and is not included in table 2.

[b]FPS ammunition data represents ammunition purchases for the federal workforce, not the contracted guard staff. FPS does not purchase ammunition for its approximately 13,500 contracted guard staff.

The amount of ammunition each component purchases for its firearm-carrying personnel also varies. We analyzed DHS data on ammunition purchases and the size of the firearm-carrying workforce for fiscal years 2008 through 2013 and found the average number of rounds of ammunition purchased per year per firearm-carrying agent or officer by component for this time period ranged between approximately 1,000 and 2,000 rounds, as shown in table 2. This variation exists because each component independently decides, based on its operational needs, how much ammunition to allocate to its firearm-carrying personnel for training and qualification each year. For example, FPS provides each officer 250 rounds per quarter per handgun for firearm qualification, while ICE provides 100 rounds per quarter per handgun for firearm qualification.

Training and Qualification Are the Biggest Drivers of DHS's Ammunition Purchase Decisions; Few Rounds Are Actually Fired in the Line of Duty

When determining annual ammunition requirements, the primary consideration for DHS components is the amount of ammunition needed to support the training and qualification of the firearm-carrying workforce, according to DHS officials. Training and qualification requirements vary for the components in our review, as do the number of rounds of ammunition typically used for training and qualification purposes. However, for all the components in our review, most firearm-carrying personnel are required to qualify four times per year on their issued firearms. We analyzed available data on DHS's ammunition requirements for the 6-year period from fiscal years 2008 through 2013. We found that the ammunition purchased by the components was reasonable given their annual expected training and operational needs.[14]

DHS Qualification and Training Requirements

DHS's more than 70,000 firearm-carrying personnel have qualification requirements they must fulfill to ensure firearms proficiency. Failure to qualify on firearms may result in the denial, suspension or revocation of credentials to carry firearms.[15] Depending on how rigorous a component's qualification requirements are and how many firearms an officer is authorized to carry, each DHS component may require a larger or smaller supply of ammunition to conduct its qualifications. Officers generally carry one or more firearms and must qualify on each firearm they carry. For example, a component may allow an officer to carry a personally owned firearm in addition to his or her duty-issued firearm, and the officer must qualify on both weapons, using ammunition provided by the relevant DHS component. DHS law enforcement officers across the department are typically required to perform quarterly firearms qualifications on their issued weapons (e.g., handgun), and in some cases demonstrate familiarity and proficiency on long guns, such as rifles and shotguns, even if they are not required to carry a long gun as part of their duties.[16] For some officers, depending on the DHS component and the requirements of their unit, more frequent qualifications or advanced firearms training is required. For example, advanced firearms training may include "down and disabled" firearms training or reduced light firearms training, among others.[17] DHS components also may have specialized units, such as tactical teams, that may require hundreds or thousands of rounds of ammunition during training activities. For example, CBP's Border Patrol Tactical Unit and Special Response Team—which enhance field operations with specialized tactics and techniques—use specialized firearms that are not otherwise issued to other CBP firearm-carrying staff. Officers in these teams are required to qualify on these additional specialized weapons, and can use more ammunition during training activities, according to CBP officials.

Along with the other five DHS components, FLETC's purchases of ammunition are to support the training of law enforcement officials to help them fulfill their responsibilities safely and proficiently. From fiscal year 2008 through 2013, FLETC has trained over 398,000 federal, state, local, tribal, and international law enforcement personnel, according to FLETC data—approximately 66,000 on average per year. FLETC provides the ammunition for the training courses provided to these law enforcement officers. FLETC officials said they determine their annual ammunition requirements based on the projected number of classes, students, and ammunition needs per class.[18] According to FLETC officials, fluctuations in FLETC's ammunition needs occur throughout the year, as a result of canceled classes, added classes, and varying numbers of students.

DHS Data Show Few Rounds of Ammunition Fired by Officers in the Line of Duty

An officer's training and proficiency on issued firearms help to ensure the safety of the officer and the public, should the officer require the use of deadly force in the line of duty. Following firearms qualification, DHS's more than 70,000 firearm-carrying personnel are issued ammunition to carry while on duty.[19] Although these officers are required by DHS to carry ammunition for their firearms, very few rounds of ammunition are actually fired in the line of duty.[20] According to DHS data, from fiscal year 2008 through fiscal year 2012, all DHS components reported firing a total of 862 rounds of ammunition in the line of duty.[21] In fiscal year 2013, all DHS components reported a combined total of 331 rounds of ammunition fired in the course of their duties compared with their total combined ammunition purchases of over 84 million rounds for that year.[22]

Additional Factors That Affect Ammunition Purchase Decisions

In addition to citing training and qualification requirements, senior DHS officials cited delivery and quality issues with ammunition, changes in firearms, historical usage rates, and the amount of ammunition currently in inventory as other factors they consider when making ammunition purchase decisions.

Delivery and Ammunition Quality Issues

Senior DHS component officials told us that another factor they consider when determining ammunition requirements is the time lag due to the ammunition delivery and quality testing process. Officials from all DHS components within our review reported time lags between placing orders for ammunition and receiving them. For example, the time lag can range from 3 to 18 months. Senior DHS component officials said that the Department of Defense gets first priority for ammunition orders from manufacturers, and supplying the military can delay orders for DHS and other federal law enforcement agencies, not only because of military priority, but also because production of other types of ammunition may be halted while manufacturers work to produce the ammunition needed by the Department of Defense. These officials also stated that while the ammunition they purchase comes from the same manufacturers that provide ammunition for commercial supply, the ammunition DHS purchases is manufactured according to DHS contract specifications, and is higher quality than commercial off-the-shelf ammunition. Senior component officials said they take this into account when determining how much and when to order ammunition, including when ammunition inventories need to be replenished. While awaiting delivery of ordered ammunition, DHS components reported using their existing ammunition inventories to meet mission requirements, such as for training and qualification.

Ammunition quality and testing are also factors in developing annual ammunition requirements for DHS. Components, such as USSS, ICE, and FLETC, have testing facilities to help ensure the ammunition ordered meets the specifications outlined in contracts with manufacturers. For example, ICE's National Firearms Tactical and Training Unit (NFTTU)— which purchases ammunition on behalf of ICE, CBP, and FPS, and provides technical assistance to TSA—operates a ballistics laboratory for the testing of ammunition, among other purposes. This unit conducts quality tests on a sample of each ammunition order before approving the acceptance and distribution of the total ammunition order.[23] According to the Assistant Director of ICE's NFTTU, if the ammunition tested does not meet NFTTU's quality standards, the batch is returned to the manufacturer and the total order is not accepted and approved until the ammunition meets standards. According to NFTTU data, this has occurred 17 times from fiscal year 2009 through fiscal year 2013, representing 3 percent of total tested batches.

Changes in Firearms and Ammunition Types

Another factor components consider when developing their annual ammunition requirements is changes in the types of firearms and ammunition used by components, according to senior DHS officials. When these changes occur, additional ammunition is required to ensure officers' proficiency on the new firearm or ammunition. For example, in August 2011, ICE management made changes to the types of authorized weapons officers could carry as their primary weapon—expanding the list of authorized 9 millimeter (mm) firearms. Prior to this change, the standard duty firearm was a .40 caliber firearm. After authorizing more 9 mm firearms, ICE experienced a 60 percent increase in the amount of 9 mm ammunition consumed, according to ICE, which resulted in ICE's inventory of 9 mm ammunition to be depleted. According to ICE officials, ICE needed to amend the existing contract for 9 mm ammunition to allow for additional purchases of ammunition to ensure officers who chose to use those 9 mm firearms had sufficient ammunition for firearms qualification and operations. As this change came at the end of fiscal year 2011, purchases increased in fiscal year 2012, as previously shown in figure 3.

Historical Usage Rates

Historical usage rates—that is, the amount of ammunition used in previous years—is also a factor components consider to determine how much ammunition to purchase, according to senior DHS component officials. DHS component officials told us they estimate how much ammunition they have used by reviewing how much ammunition they purchased in previous years, the number of authorized firearm-carrying personnel, and related training and qualification requirements.[24] According to our analysis of estimated ammunition usage data provided by DHS components, for fiscal years 2008 through 2013, DHS components in our review estimated using, on average, approximately 110 million rounds of ammunition per fiscal year, with a high of about 141 million rounds in fiscal year 2009 to a low of about 89 million rounds in fiscal year 2013 (see figure 4).

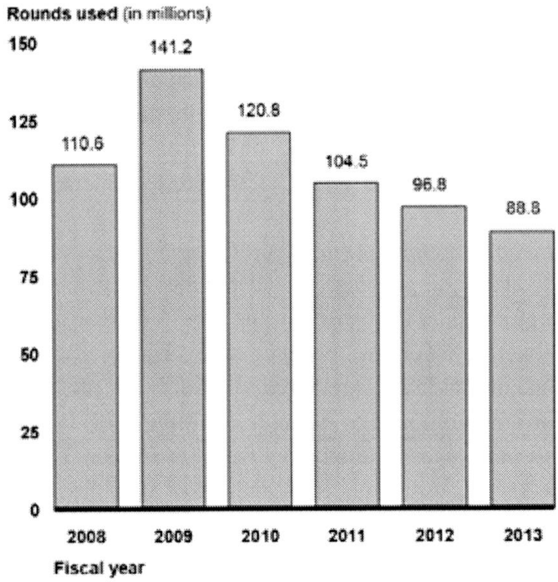

Source: GAO analysis of DHS data.

Figure 4. Estimated Ammunition Used by the Department of Homeland Security (DHS) Components in Our Review from Fiscal Years 2008 through 2013.

Ammunition Inventory

The amount of ammunition components have in inventory is also a factor to determine how much and when to order more ammunition, according to DHS component officials. DHS components determine how much ammunition to keep in inventory, and when to replenish ammunition inventories. The six DHS components in our review stated they strive to maintain about 12 to 24 months' of ammunition inventory to meet the training, qualification, and operational ammunition needs of firearm-carrying personnel. As stated earlier, there can be months-long delays between placing an order for ammunition and receiving it. To help ensure components have sufficient ammunition on hand to support the training and operational needs of their officers, DHS components maintain inventories of ammunition. For example, TSA officials said they try to keep a large ammunition inventory on hand, because the FFDOs (trained and armed pilots) are allowed to provide as little as 24 hours notice to perform their qualification, based on their fluctuating schedules.

Ammunition is stored in limited supply in the field offices of the various components, as DHS component officials have said there is limited storage for ammunition in the field. Officials added that while ammunition is typically shipped directly from the manufacturers to the field offices that placed the orders, there are occasions when production delays require components to ship ammunition from their reserves stored in armories to resupply the field. It is important to note that inventory levels reflect a point in time and that inventory fluctuates throughout the year as ammunition orders arrive from manufacturers and ammunition is used for firearms qualification and other purposes in the field. Therefore, it is not possible to provide an average ammunition inventory for a year or past years, because the inventory is constantly fluctuating. Table 3 shows how much ammunition DHS components had in inventory (in both estimated rounds and approximate number of months' supply) at three different points in time between November 2012 and October 2013, as reported to us by DHS components. We provide these estimates at three points in time to show, by component, how ammunition inventory fluctuates over the course of 1 year.

Table 3. Selected Department of Homeland Security (DHS) Components' Estimated Ammunition Inventory by Total Rounds and by the Number of Months' Supply at Three Points in Time Over 1 Year

DHS component	Estimated rounds in inventory, November 2012	Estimated rounds in inventory, April 2013	Estimated rounds in inventory, October 2013
U.S. Customs and Border Protection	94 million (32 months)	94 million (33 months)	72 million (25 months)
Immigration and Customs Enforcement	42 million (21 months)	29 million (14 months)	40 million (20 months)
Secret Service	6 million (17 months)	6 million (18 months)	7 million (21 months)

Table 3. (Continued)

DHS component	Estimated rounds in inventory, November 2012	Estimated rounds in inventory, April 2013	Estimated rounds in inventory, October 2013
Transportation Security Administration[a]	30 million	27 million	14 million
Federal Protective Service	3 million (15 months)	3 million (16 months)	3 million (17 months)
Federal Law Enforcement Training Centers	19 million (18 months)	17 million (16 months)	23 million (21 months)
DHS total	193 million (26 months)	178 million (24 months)	159 million (22 months)

Source: GAO analysis of DHS data.

Notes: To determine approximate number of months' supply of ammunition in inventory for each component, we used fiscal year 2013 estimated ammunition usage data provided by DHS components and divided that number by 12 to get an estimate of monthly ammunition usage. We then divided the components' inventory figures by the monthly ammunition usage estimate to get estimated months' worth in inventory for those three points in time. For more information on our analysis, see appendix I.

In 2013, CBP and TSA made changes to their inventory reporting protocols, which officials said can explain the changes in inventory between the dates.

DHS components' records may not always reflect the exact quantities of the ammunition inventory at a given time. For example, because inventory data may not be updated in real time, quantities of ammunition actually available and those reflected on the inventory records may differ. To assess the reliability of these data, we reviewed the extent to which the agencies have procedures and controls for ensuring that the data are consistent and accurate, and interviewed knowledgeable officials responsible for collecting and reporting the data. We determined the data to be sufficiently reliable to be reported rounded to the millions. To maintain an acceptable level of accuracy in inventory records, DHS officials said that DHS and the components have established inventory control policies and procedures, including taking physical inventories and conducting audits to verify ammunition balances, as discussed later in this report.

[a]The approximate number of months' supply for TSA is considered sensitive security information and is not included in table 3.

Each component is responsible for maintaining and recording its own ammunition inventory, and, according to senior DHS component officials, ammunition balances may not always be up to date and generally are estimates. For example, a field location may show more ammunition in inventory than other field offices, but that could reflect that the officers have not yet completed a quarterly firearms qualification or the inventory tracking system has not yet been updated to reflect changes. In addition, components may also consider ammunition that has been ordered, but not yet arrived, to be part of their "current" inventory. For example, TSA reported an inventory level of almost 30 million rounds in November 2012, but TSA officials said that number included ammunition orders that had been placed with the manufacturer, but were not yet in physical inventory. According to TSA, the quantity of ammunition in TSA's physical possession at the time of the inquiry was approximately 19 million rounds.

DOJ Components Cited Similar Factors in Determining Ammunition Requirements for Personnel Authorized to Carry Firearms and Annual Ammunition Purchases are Comparable to Those of DHS

Similar to DHS, the DOJ components in our review also purchase ammunition to ensure their personnel authorized to carry firearms are equipped to carry out their various law enforcement missions. For the three DOJ components in our review, ammunition purchases are driven primarily by the firearms training and qualification requirements for their firearm-carrying workforce. According to DOJ data, DOJ component firearm qualification requirements range from semi-annual to quarterly, depending on the component; however, most require quarterly firearms qualification on an officer's primary weapon. As shown in table 4, the approximate number of rounds of ammunition purchased per firearm-carrying agent or officer for the three DOJ law enforcement agencies in our review for fiscal years 2011 through 2013 were similar to the amounts for DHS components in our review.[25] For DOJ components, the average number of rounds of ammunition purchased per authorized firearm-carrying agent or officer per year across fiscal years 2011 through 2013 was approximately 1,300 rounds. For DHS, the average number of rounds of ammunition purchased per authorized firearm-carrying agent or officer per year for fiscal years 2011 through 2013 was approximately 1,000 rounds. Senior DOJ officials also reported similar considerations when making ammunition purchases, such as ammunition delivery lead time, quality issues, and the amount of ammunition in inventory. As with DHS officials, DOJ officials reported that the type of ammunition they use is manufactured to a higher standard than ammunition purchased by the civilian market and is not considered off-the-shelf.

Table 4. Approximate Number of Rounds of Ammunition Purchased per Agent or Officer for the Department of Homeland Security and Department of Justice Components in Our Review for Fiscal Years 2011 through 2013

	2011	2012	2013
Number of rounds purchased per agent or officer at the Department of Homeland Security	1,100	1,200	900
Number of rounds purchased per agent or officer at the Department of Justice	1,400	1,300	1,200

Source: GAO analysis of Department of Homeland Security and Department of Justice data.

According to DOJ component officials, the ammunition purchased by DOJ is produced after an order is placed with a manufacturer, and the time between placing an order and receiving it can be several months. Quality testing of ammunition is also part of DOJ components' procurement process, to help ensure ammunition performs as expected in the field. Additionally, DOJ components also consider the amount of inventory on hand when determining ammunition requirements. According to officials from the DOJ components in our review, the amount of inventory components try to maintain ranged from a minimum of 6 months to a maximum of 24 months. Ammunition inventory data provided by two of the three DOJ components indicated that inventory ranged from about 13 months' worth to about 20 months' worth.[26] This range is similar to the range reported by DHS components.

DHS Has Policies and Guidance for Firearms and Ammunition Inventory Management

DHS Components Have Inventory Management Policies and Procedures and Are in the Process of Implementing Revised DHS Firearms Asset Policy

Each of the DHS components in our review has policies, procedures, and processes that describe the requirements and guidance necessary to ensure inventory management, control and accountability for firearms and ammunition.[27] These policies include describing who is responsible and accountable for firearms and ammunition, tracking and accounting for firearms and ammunition throughout their life cycles, and periodically conducting physical inventories to verify the existence and location of the firearms and ammunition recorded in the accountability records. For example, FLETC requires the firearms and weapons custodian to perform a daily audit of all ammunition issues and receipts occurring at specified issue points. TSA requires supervisors to conduct unannounced employee firearm inspections and recommends that TSA offices conduct random, selective, periodic, or unannounced internal audits to ensure proper inventory controls and accountability. In addition, ICE, CBP, and FPS use an automated system called Firearms, Armor, and Credentials Tracking System (FACTS) to provide visibility of and oversight over their firearms inventories. We did not assess if the components' policies, procedures, and process for managing their firearms and ammunition inventories were working as intended.

In 2010, the DHS Inspector General reported that the department's management and oversight of components' safeguards and controls over firearms were not effective, because DHS did not have specific firearms policies and procedures in place. [28] The report noted that DHS relied on components to modify existing policies and procedures with specific guidance for controlling firearms, but component personnel did not always follow component policies, resulting in a loss of firearms. Further, according to that report, DHS did not require independent third parties to conduct annual firearm inventories to ensure that components were accountable for firearms. Consequently, the DHS Inspector General report recommended, among other things, that DHS develop department-wide policies and procedures for safeguarding and controlling firearms.[29]

In response to the findings and recommendations of the report, in 2012 DHS issued the DHS Firearm Asset Policy (firearm policy) to govern the components' management of firearms, including inventory management requirements. According to the firearm policy, components are responsible for developing their own controls and for overall management of their respective firearms programs, which includes adhering to DHS requirements for firearms management. The components are to integrate the firearm policy into their operations during fiscal year 2013, with a planned full implementation by the end of fiscal year 2014. Among other things, DHS's firearms policy requires five specific inventory control measures, noted below:

- **Establish policies and guidance:** Components are responsible for establishing policies and guidelines to ensure that the firearm asset management system of record is updated throughout the asset life cycle to document all transactions and events.

- **Conduct annual physical inventory:** The component's property management officer or firearm program manager is to ensure that the component's inventory plan incorporates an annual physical inventory for all firearms. For internal control purposes, there are to be at least two individuals conducting a firearm physical inventory. Data on every firearm shall be recorded and maintained in the firearm asset management system of record.[30]
- **Conduct independent third party audits:** Components or DHS's Chief Readiness Support Officer's Office are to engage independent third parties to conduct an annual audit of at least 15 percent of their firearm inventory.
- **Ensure inventory accuracy:** Supervisors are to conduct unannounced inventory verifications to ensure accuracy of firearms inventory.
- **Verify issued firearms:** In addition to an annual physical inventory, each component's property management officers or firearm program managers are to ensure quarterly firearm inventory verifications are conducted for all firearms issued.[31]

For the six DHS components in our review, we found that each of the components established policies and guidelines to manage firearms assets, and some of the components were already implementing aspects of the new firearm policy.[32] For example, all six components' policies require that they conduct a complete annual physical inventory of all firearms. Two of the components' policies require them to engage an independent third party to conduct an annual audit of at least 15 percent of their firearm inventories. In addition, three of the six components have procedures in place to ensure quarterly firearm inventory verifications are conducted for all firearms issued. DHS officials said that components are not expected to address all of the requirements in the DHS firearm policy until the end of fiscal year 2014. Therefore, it is too early to know if components will meet all of the requirements within this time frame. Component officials we spoke with said that the implementation time frame may require additional time to work through the feasibility of certain requirements. For example, ICE officials noted that components with collective bargaining units, such as CBP and ICE, for example, would need agreement by the collective bargaining unit representing a component's law enforcement personnel before certain aspects of the new firearm policy can be implemented.

DHS Components Have Policies and Procedures for Managing Ammunition Inventory

DHS issued a manual in 2013 on personal property asset management that provides a general description of controls for managing property, including ammunition, which is considered a sensitive asset requiring certain controls.[33] Specifically, with respect to inventory management of ammunition, the DHS manual requires that ammunition be physically inventoried at least annually. In accordance with the DHS property management manual, DHS relies on the components to establish specific policies and procedures for managing, safeguarding, and controlling their ammunition inventories. We found that all six DHS components in our review have policies and procedures for managing their ammunition inventories, including requiring that a physical inventory be conducted at least annually, which is in accordance with the DHS property management manual. For example, USSS guidance requires that field office personnel conduct annual inventories to ensure the completeness and accuracy of the ammunition inventory information within their inventory system.

Officials from five of the six components reported that they do conduct physical inventories of their ammunition at least annually. For example, FLETC officials reported that personnel conduct an annual inventory in which they are to reconcile ammunition with control records through a dual verification process to ensure accuracy by performing a physical and visual verification. One component—ICE—reported that it does not conduct a distinct physical inventory of ammunition (that is, at one point in time reconcile ammunition on hand with ammunition expected to be in inventory). However, ICE officials stated that ammunition inventories are conducted by ICE field-level officials at various times during the year for determination of needs. According to ICE officials, ICE has treated ammunition as a consumable asset and has not subjected it to the same inventory process as firearms.[34] Sensitive assets such as firearms are subject to the special requirements outlined in the DHS property management directive. ICE officials stated that although they have not conducted annual physical inventories, ammunition balances and inventory levels are continuously tracked in FACTS, and senior firearms instructors in local field offices manage and maintain their ammunition inventory in FACTS as well as on ammunition inventory control sheets that are maintained locally. ICE officials said beginning in fiscal year 2014, they plan to institute a separate and distinct physical inventory of ammunition in conjunction with the annual ICE-wide sensitive asset inventory.

AGENCY COMMENTS

We provided a draft of this report to DHS and DOJ for official review and comment. DHS provided written comments. DHS agreed that annual ammunition purchases have declined during the past 3 years and are comparable to those of DOJ law enforcement agencies. DOJ did not provide written comments on this report. DHS and DOJ provided technical comments, which we incorporated where appropriate.

David C. Maurer
Director, Homeland Security and Justice Issues

APPENDIX I: OBJECTIVES, SCOPE, AND METHODOLOGY

1) What are the trends in the Department of Homeland Security's (DHS) ammunition purchases since fiscal year 2008, what factors affect its purchase decisions, and how do DHS's purchases compare with those of the Department of Justice's (DOJ)?
2) What policies and guidance does DHS have for managing firearms and ammunition inventories?

To obtain information on DHS's acquisition practices for ammunition, we examined the Federal Acquisition Regulation (FAR); DHS's acquisition and procurement policies and guidance, including the Homeland Security Acquisition Regulation (HSAR); and the *Homeland Security Acquisition Manual* (HSAM). We reviewed prior GAO reports and a Congressional Research Service memorandum on DHS acquisition and procurement.[1] We also reviewed contracts for ammunition purchases and related pre-award documentation, including acquisition plans, market research, and cost estimates, but we did not review whether components' ammunition procurements complied with the FAR, HSAR, or HSAM. We interviewed the

Director for Procurement Policy and Oversight, the Director for Oversight and Strategic Sourcing, and the acquisition and program officials from six DHS components including U.S. Customs and Border Protection (CBP), U.S. Immigration and Customs Enforcement (ICE), Transportation Security Administration (TSA), U. S. Secret Service (USSS), National Protection and Programs Directorate/Federal Protective Service (FPS), and Federal Law Enforcement Training Centers (FLETC) to discuss contracting procedures and practices for procuring ammunition, including the use of the Weapons Ammunition Commodity Council and strategic sourcing to coordinate and leverage their buying power. We reviewed relevant policies and guidance, examined department and component documentation, and interviewed officials from the DHS Office of the Chief Procurement Officer and components on practices they employ and contracts they have leveraged. Specifically, we interviewed officials from DHS's strategic sourcing program office, as well as all six components, to gain an understanding of DHS's strategic sourcing processes, the availability of strategic sourcing contracting vehicles for ammunition, and the potential of establishing component-specific strategic sourcing goals for procuring ammunition. We also interviewed Office of the Chief Procurement Officer Oversight and Pricing Branch officials regarding their tri-annual reviews of procurement operations.

To determine any trends in DHS's ammunition purchases since fiscal year 2008, we obtained available data from DHS law enforcement components with firearm-carrying personnel regarding their ammunition purchases, costs, usage, and the size of the authorized firearm-carrying workforce for fiscal years 2008 through 2013. Specifically, we selected all DHS law enforcement components with firearm-carrying personnel— CBP, ICE, TSA, USSS, FPS— as well as FLETC. The number of TSA federal air marshals and the number of federal flight deck officers authorized by TSA to carry firearms is considered sensitive security information. Accordingly, we excluded that information from this report. We also obtained data from DHS components on their estimated ammunition inventory balances for three points in time—November 2012, April 2013, and October 2013. Because we determined that it is not possible to provide an average ammunition inventory for a year or past years, as ammunition inventory is constantly fluctuating, we selected these points in time based on data that was previously reported by DHS to Congress prior to our review as well as data available at the start and end of our review. In addition, DHS components' records may not always reflect the exact quantities of the ammunition inventory. For example, depending on how often components update their inventory data, the time lag may result in differences between quantities of ammunition actually available and those reflected on the inventory records. We also obtained data on planned ammunition purchases for fiscal year 2014. In deciding which DHS components to examine in our review, we excluded components that were not law enforcement components and had small numbers of personnel authorized to carry firearms and made small purchases of ammunition, such as the Office of the Inspector General and the Federal Emergency Management Agency (FEMA). We also excluded the U.S. Coast Guard—a DHS component with firearm-carrying personnel—because the Coast Guard does not procure ammunition through DHS; rather, the U.S. Coast Guard procures its ammunition through Department of Defense contracts. Similarly, we selected all DOJ law enforcement components with firearm-carrying personnel regarding their ammunition purchases, costs, usage, and size of the workforce authorized to carry firearms for fiscal years 2011 through 2013. The specific DOJ law enforcement components included in our review are the Bureau of Alcohol, Tobacco, Firearms and Explosives (ATF), Federal Bureau of Investigation (FBI), and U.S. Marshals Service. The Federal Bureau of Prisons could not provide ammunition purchase data comparable to that of the other components based on its method of record keeping and data retention policy. We determined that data from the Drug Enforcement Administration (DEA) were not sufficiently reliable for our purposes, because DEA could only provide estimates of purchases. We, therefore, excluded both agencies from our scope. We included the selected DOJ law enforcement agencies in our review to provide perspective and context to help understand DHS's ammunition purchases and usage relative to those of other federal law enforcement agencies.[2]

We selected fiscal year 2008 as the starting point for our analysis because data on ammunition purchases, cost, and use were not readily available or reliable prior to fiscal year 2008, according to DHS officials. Nevertheless, obtaining data going back 5 fiscal years provides some historical perspective on DHS's ammunition purchases and use over time, and allows for comparisons with ammunition purchases for fiscal year 2013 and planned purchases for fiscal year 2014. We analyzed the data to identify any changes in the amount of ammunition purchased or used year to year, and we interviewed knowledgeable officials from the components to understand some of the reasons why changes occurred. For each fiscal year, we calculated the amount of ammunition purchased per officer authorized to carry firearms at each DHS and DOJ component in our review, using component-provided data on the amount of ammunition purchased and the number of officers. We calculated the average number of rounds purchased per agent or officer per year for fiscal years 2008 through 2013 at DHS by determining the average ammunition purchased across components in each fiscal year, excluding FLETC, and dividing that by the average number of officers at DHS for each fiscal year.[3] We determined that data for DOJ agencies were not readily available prior to fiscal year 2011 for all components, so we used data for fiscal years 2011 through 2013 to do the same calculation. We analyzed the amount of ammunition purchased per agent or officer authorized to carry a firearm to account for the varying sizes of the different departments and components. We selected DOJ to provide a comparison with DHS's ammunition purchases as it has the second largest number of personnel authorized to carry firearms. However, our DOJ analysis is not to suggest that the department represents a model or standard against which DHS is assessed. Differences in the amount of ammunition per firearm-carrying personnel reflect a number of factors, including unique mission requirements and training needs.

To assess the reliability of the data we obtained from DHS components on ammunition purchases and costs, we reviewed the extent to which the components have procedures and controls for ensuring that the data are consistent and accurate, and interviewed knowledgeable officials responsible for collecting and reporting the data. In addition, for the data we obtained from DHS components on ammunition purchases and costs, we compared it with corroborating evidence to determine data consistency and reasonableness. Specifically, we obtained information on ammunition procurements using obligations data from fiscal years 2008 through 2012, and part of fiscal year 2013 from the Federal Procurement Data System (FPDS), which tracks all contracts using appropriated funds government-wide. We used the obligations data from the FPDS and compared ammunitions obligations for DHS with the sum of ammunition costs as provided to us by each of the components. We also compared estimated ammunition used with ammunition purchased under the assumption that in a given year, components would purchase ammunition to replace what was used. We calculated how much ammunition DHS components might be expected to use in a given year using data, provided by DHS, on the size of the firearm-carrying workforce, the number of staff qualified on various types of firearms, and the number of rounds of ammunition typically allotted for training and qualifications. We also reviewed selected DHS ammunition procurement contracts, which were readily available, for information on ammunition purchases, costs, and stated requirements. We found the data on ammunition purchases and costs for the DHS components to be sufficiently reliable for our purposes. To assess the reliability of the data we obtained from DHS components on ammunition inventory, we reviewed the extent to which the agencies have procedures and controls for ensuring that the data are consistent and accurate, and interviewed knowledgeable officials responsible for collecting and reporting the data. To determine the approximate number of months' supply of ammunition in inventory for each component, we used fiscal year 2013 estimated ammunition usage data provided by DHS components and divided that number by 12 to get an estimate of monthly ammunition usage. We then divided the components' inventory figures by the monthly ammunition usage estimate to get estimated months' worth in inventory for those three points in time. Each component has policies in place to maintain and record its own ammunition inventory, but according to officials, ammunition balances may not always be up to date and generally are estimates. For example, depending on how often components update their inventory data, the time lag may result in differences between quantities of ammunition actually available and those reflected on the inventory records. On the basis of this information, we determined the data to be sufficiently reliable to be reported rounded to the millions.

To assess the reliability of the data we obtained from DHS components on ammunition usage, we interviewed knowledgeable officials responsible for collecting and reporting the data. DHS components do not centrally track historical ammunition usage, in large part because their data management systems are not designed to track ammunition usage. However, components did provide us estimates of ammunition usage for fiscal years 2008 through 2013 based on how much ammunition they purchased in previous years, the number of firearm-carrying personnel, and related training and qualification requirements. We determined these data were sufficiently reliable to report general trends in usage when rounded to the hundred thousands.

DHS components provided us with data on the number of personnel authorized to carry firearms for fiscal year 2008 through 2013 and information about how they compiled these data. Data sources varied by component and included personnel systems, the Firearms, Armor, and Credentials Tracking System (FACTS), and data from the National Finance Center. Workforce data provided by CBP is not precise because it included a rounded number of personnel for one of CBP's subcomponents. Therefore, we report CBP's data rounded to the thousands. On the basis of this information, we determined these data were sufficiently reliable to be used for background purposes and in our calculations of the average number of rounds purchased per agent or officer. To assess the reliability of the inventory, workforce, and ammunition purchase data we obtained from DOJ, we interviewed knowledgeable officials responsible for collecting and reporting the data about the extent to which DOJ components have procedures and controls in place for ensuring that the data are consistent and accurate. We calculated the DOJ inventory months' worth estimates in the same way we did for DHS components. On the basis of this information, we determined these data to be sufficiently reliable to report rounded estimates based on these data. We interviewed officials from the six DHS components in our review to discuss the ammunition procurement process and how ammunition requirements are determined at the component level and the purpose of quality testing the ammunition, and to understand the context of why proficiency among firearm-carrying personnel is important. In addition, DHS components provided examples of purchase and delivery orders in which the components experienced a variation in lag time between placing a purchase order and receiving a shipment of ammunition, which can result from the manufacturing lead time associated with ammunition purchases. Finally, for additional context, we conducted a site visit to the ICE National Firearms Testing Lab, which performs quality control testing on ammunition and firearms, and maintains an inventory of ammunition and firearms for distribution to DHS field locations. To address our second question, we examined the general requirements and policies DHS and its components had to manage and oversee firearms and ammunition, but we did not assess whether the directives, policies, and guidance were working as intended or if component personnel were adhering to them, as that was outside the scope of our review. We examined agency-wide directives and guidance, and component management policies and procedures for managing ammunition and firearms. This included reviewing the firearm policy manuals or related documentation from CBP, FLETC, FPS, ICE, TSA, and USSS to determine whether the manuals specifically addressed agency-wide firearm and ammunition inventory management requirements. We also interviewed officials from the Office of the Chief Readiness Support Officer and from DHS components, and reviewed components' written responses regarding their policies and guidance for managing firearms and ammunition. In addition, we interviewed component officials responsible for conducting management and compliance reviews of component operations, including on-site review and self-inspection program findings and recommendations, which include, but are not limited to, ammunition and firearms management and controls. We also reviewed a 2010 DHS Inspector General report on DHS firearms controls.[4]

We conducted this performance audit from May 2013 to January 2014 in accordance with generally accepted government auditing standards. Those standards require that we plan and perform the audit to obtain sufficient, appropriate evidence to provide a reasonable basis for our findings and conclusions based on our audit objectives. We believe that the evidence obtained provides a reasonable basis for our findings and conclusions based on our audit objectives.

APPENDIX II: COMMONLY USED AMMUNITION AND ISSUED FIREARMS AT DHS

Figure 5 shows some of the most commonly purchased types of ammunition for ICE, and figure 6 shows firearms commonly issued by DHS components.[1]

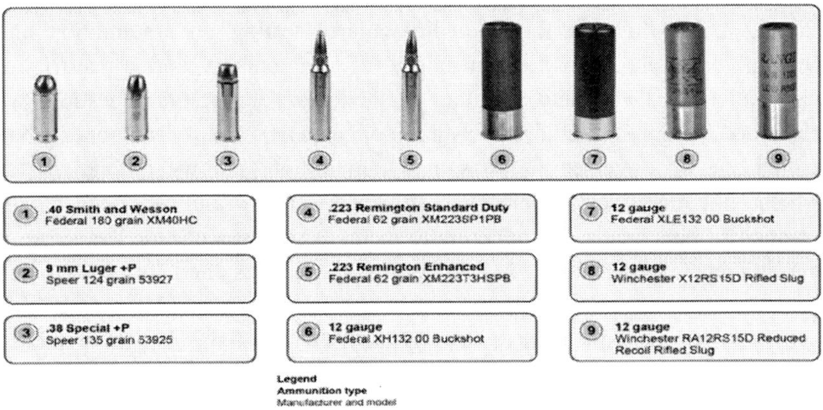

Source: GAO analysis of Department of Homeland Security data.

Figure 5. Ammunition Commonly Purchased by Immigration and Customs Enforcement (ICE).

Source: GAO analysis of Department of Homeland Security data.

Note: The firearms illustrated are intended to be general representations and are not specifically the model types or configurations used by DHS components. DHS components may use other weapons in addition to the ones in this figure.

Figure 6. Firearms Commonly Issued by Department of Homeland Security (DHS) Components.

Appendix III: Department of Homeland Security Ammunition Contracts, as of October 1, 2013

Most of DHS's ammunition contracts, whether strategically sourced or individual contracts, are indefinite delivery, indefinite quantity (IDIQ) contracts, which are typically negotiated for a base year with additional options for purchasing ammunition up to a certain maximum number of rounds, or contract ceiling. As shown in the table below, the 29 existing DHS ammunition contracts extend over the next 4 fiscal years and have a remaining contract limit of approximately 704 million rounds (for various ammunition types) if every option for purchasing ammunition were exercised into fiscal year 2018.

Component name	Ammunition caliber	Contract award date	Contract end date	Contract ceiling in dollars	Contract ceiling in rounds	Estimated remaining number of roundsavailable as of October 1, 2013
Federal Law Enforcement Training Centers (FLETC)	.357, .45, .40 and 9 mm	03/28/2013	4/1/2014	$30,000,000	104,587,924	According to FLETC officials, thesecontracts are Reduced Hazard TrainingAmmunition (RHTA) and are available forthe entire federal government to orderfrom. Orders may be placed against thesecontracts before 4/1/2014 and are notexpected to meet the ceiling. Althoughthey are still active, DHS intends toreplace them with new contracts.

(Continued)

Component name	Ammunition caliber	Contract award date	Contract end date	Contract ceiling in dollars	Contract ceiling in rounds	Estimated remaining number of roundsavailable as of October 1, 2013
	.357, .40, 9 mm, 12 gauge, and .223	03/29/2013	4/1/2014			
	.357, .45, .40, 9 mm, 12 gauge, and .223	03/30/2013	4/1/2014			
	.40, 10mm, 9 mm, and .223	03/31/2013	4/1/2014			
	.223, 30-06, .308, 12 gauge, .357, .38, .40, .45, 7.62, and 9 mm	04/12/2013	04/11/2018	6,229,785	23,000,000	22,992,500
	.223, 30-06, .308, 12 gauge, .357, .38, .40, .45, 7.62, and 9 mm	04/12/2013	04/11/2018	4,295,419	23,000,000	20,210,000

Component name	Ammunition caliber	Contract award date	Contract end date	Contract ceiling in dollars	Contract ceiling in rounds	Estimated remaining number of roundsavailable as of October 1, 2013
	.223, 30-06, .308, 12 gauge, .357, .38, .40, .45, 7.62, and 9 mm	04/12/2013	04/11/2018	$6,359,270	23,000,000	19,874,000
	.45, .22 , 3030, 12 gauge, .357, and 7 62	04/15/2013	12/11/2013	29,520	99,800	0
	.40	06/11/2013	11/12/2013	500,000	2,000,000	0
	.40 and .223	08/29/2013	12/13/2013	373,825	1,650,000	0
	.40	09/04/2013	11/04/2013	42,024	200,000	0
	12 gauge	09/03/2013	12/15/2013	2,765	5,000	0
	.45	09/16/2013	01/16/2014	35,555	100,000	0
	.38	09/16/2013	01/16/2014	7,950	30,000	0
Immigration and Customs Enforcement (ICE)a	9 mm	9/29/2011	9/28/2016	3,493,331	14,750,000	6,149,000
	.38	9/29/2011	9/28/2016	3,334,518	8,800,000	8,093,000
	.40	1/10/2012	1/9/2017	109,490,400	450,000,000	408,461,000
	.223	7/12/2012	7/11/2017	10,000,000	10,000,000	9,872,200

(Continued)

Component name	Ammunition caliber	Contract award date	Contract end date	Contract ceiling in dollars	Contract ceiling in rounds	Estimated remaining number of roundsavailable as of October 1, 2013
	.223	1/1/2013	12/31/2017	92,983,440	172,000,000	171,011,800
Transportation Security Administration (TSA)	.357	9/26/2013	9/25/2014	744,095	3,454,000	0
U.S. Secret Service (USSS)	Specific ammunition types are considered law enforcement sensitive and are therefore not included in this report	6/23/2010	5/29/2015	10,902,329	43,350,000	29,749,000
		6/28/2010	5/31/2015	501,690	1,250,000	305,000
		8/16/2010	5/29/2015	682,900	1,600,000	1,560,000
		6/1/2010	5/29/2015	242,327	3,750,000	690,250
		8/15/2010	5/29/2015	$876,000	800,000	217,500
		6/23/2010	5/29/2015	1,606,000	3,000,000	3,000,000
		7/14/2010	5/29/2015	1,606,000	3,000,000	1,885,000
		9/30/2013	9/29/2018	750,000	400,000	320,000

Component name	Ammunition caliber	Contract award date	Contract end date	Contract ceiling in dollars	Contract ceiling in rounds	Estimated remaining number of roundsavailable as of October 1, 2013
		8/24/2013	12/31/2013	267,500	963,000	0
Total				$ 285,356,645	894,789,724	704,390,250

Source: GAO analysis of Department of Homeland Security contract data.

Notes: Totals may not add up because of rounding.

DHS's strategic sourcing contract vehicles include contracts or agreements that have been established for use by two or more components to leverage their buying power and secure competitive prices resulting in cost savings through collective procurement actions (e.g., buying in bulk quantities at lower prices). Strategically sourced contracts are shaded.

[a]Other components purchase on ICE contracts, including U.S. Customs and Border Protection (CBP) and Federal Protective Service (FPS).

End Notes

[1] DHS defines a firearm as any weapon (excluding flare and starter guns) that will, or that is designed to, or may be readily converted to expel a projectile by the action of an explosive to include the frame or receiver of any such weapon or any sound suppression device. DHS classifies a firearm as a "Dangerous and Hazardous Sensitive Asset."

[2] The number presented here—more than 70,000—represents the number of DHS personnel required to carry a firearm, according to data provided by DHS, as of October 2013. The specific number of DHS personnel required to carry a firearm is considered sensitive security information and therefore is not included in this report. This is because the specific number of DHS personnel would include the number of TSA firearm-carrying personnel, such as federal air marshals and federal flight deck officers—and those numbers are considered sensitive security information. In addition to these more than 70,000 civilian DHS law enforcement officers, the U.S. Coast Guard—a DHS component—has over 41,000 uniformed members of the military who also train with and carry firearms in the performance of their duties.

[3] While other DHS components, such as the Office of the Inspector General and the Federal Emergency Management Agency, may have a small number of personnel authorized to carry firearms for whom they purchase ammunition, these components are not law enforcement components, and thus we did not include them in our review. The number of TSA firearm-carrying personnel is considered sensitive security information. Accordingly, some information on TSA ammunition purchases and usage that could identify the number of TSA firearm-carrying personnel was excluded from this report.

[4] Commonly used firearms and ammunition refer to the firearms issued and largest amounts of ammunition purchased by the components in our review.

[5] DHS defines strategic sourcing as a structured process of analyzing DHS spending and using a collaborative approach to make business decisions about acquiring and managing commodities and services more effectively and efficiently across multiple components or the entire department.

[6] Among other things, we reported that DHS had the most initiatives and had the highest percentage of spending through strategic sourcing of the four selected agencies we reviewed. GAO, *Strategic Sourcing: Improved and Expanded Use Could Save Billions in Annual Procurement Costs*, GAO-12-919 (Washington, D.C., Sept. 20, 2012).

[7] Prior to August 2012, strategic sourcing contract vehicle use was encouraged by DHS, but not required.

[8] DHS estimated savings based on past government orders and open market pricing when available. Starting in fiscal year 2013, DHS officials said that DHS strategic sourcing contracts will calculate savings based on market research and past government orders now that information is more readily available.

[9] Qualification training involves the safe handling, proficient application, and justifiable use of firearms to help ensure a law enforcement officer's ability to safely handle required weapons.

[10] The selected DOJ components are FBI, ATF, and U.S. Marshals Service. Data on DOJ ammunition purchases prior to fiscal year 2011 were not readily available; therefore, we included fiscal years 2011 through 2013.

[11] The National Shooting Sports Foundation is a trade association for the firearms industry, representing more than 8,000 manufacturers, distributors, firearms retailers, shooting ranges, sportsmen's organizations and publishers.

[12] We excluded FLETC from our per officer analysis because FLETC has few firearm-carrying officers for which it purchases ammunition. Ammunition purchased by FLETC is utilized by all students who train on firearms at its facilities, including DHS and other federal law enforcement personnel, as well as state, local, tribal, and international law enforcement personnel.

[13] FPS ammunition data represent ammunition purchases for the federal FPS workforce, not the contracted guard staff. FPS does not provide ammunition for its approximately 13,500 contracted guard staff.

[14] To conduct this analysis, we estimated how much ammunition DHS components might be expected to use in a given year by considering the size of the firearm-carrying workforce, the number of agents qualified on how many types of firearms, and the amount of ammunition typically allotted for each firearm qualification by component. For more information on our methodology, see appendix I.

[15] Demonstrating proficiency may include displaying proper handling techniques and dexterity, such as holstering, loading, unloading, and operating the firearm; shooting with the left and right hands; demonstrating safely handling firearms and ammunition; achieving at least a minimum numerical score on courses of fire from varying distances from the target; and completing advanced firearms training exercises, among other requirements.

[16] CBP, ICE, FPS, TSA's FAMS, and USSS firearm-carrying personnel are required to qualify on firearms at least four times per year. TSA's FFDOs are required to qualify two times per year on handguns.

[17] "Down and disabled" firearms training is intended to ensure proficiency even after an officer has been wounded during an event requiring use of force.

[18] According to component officials from FLETC, USSS, and ICE, these components have explored the use of simulation technology to support law enforcement training needs. Simulated firearms training technology may provide benefits to law enforcement personnel during basic marksmanship training. For example, simulation training with laser-fired firearms could allow instructors to carry on normal conversations while instructing students in the proper stance, grip, sight alignment, and trigger control, since hearing protection would not be necessary. Also, instructors could get closer to students to see errors in weapon handling, especially with respect to grip and trigger press, that could not have normally been seen with live fire weapons. Other potential benefits of using simulators are reduced ammunition usage and range maintenance costs. FLETC's annual recurring costs for simulators are $73,000 per year. However, component officials have also raised concerns about the need for officers to train on live fire firearms to gain full proficiency on their weapons, since simulation training does not currently accurately replicate recoil on firearms. Component officials have also said that while there would be reduced need for ammunition, the simulator technology would also require regular maintenance and upgrades, which would also come with a cost.

[19] DHS components generally use the same type of ammunition for training and qualification as they use for duty (operational use). According to DHS, this is because delivering and storing different types of ammunition for training and operational use creates complex logistical challenges, and could create an officer safety issue if the wrong ammunition is used in the field. FLETC primarily uses Reduced Hazard Training Ammunition, which is free of lead and other toxic substances, in its firearms training curriculum.

[20] The number of rounds carried by an officer or agent on duty varies by factors such as DHS component, firearm type, and number of firearms, among other factors.

[21] According to CBP, it did not have data on ammunition fired on duty for fiscal years 2008 and 2009 because prior to January 1, 2010, CBP did not require reporting the number of rounds fired for officer-involved shootings. After January 1, 2010, CBP began recording all use of force incidents in the CBP Use of Force Reporting System, which required more detailed information regarding incidents. Because CBP data prior to January 2010 may not account for all rounds fired, fiscal years 2008 and 2009 were not included in CBP's responses.

[22] These totals may include rounds that were accidently discharged.

[23] NFTTU assists ICE by providing personnel with firearms, intermediate force weapons, protective equipment, training, guidance, as well as tactical and logistical support. The unit's mission is to increase safety and improve the tactical proficiency of the armed workforce while maintaining accountability for property.

[24] Components do not centrally track historical ammunition usage rates, in large part because their data management systems are not designed to track ammunition usage. As of 2013, the database used by ICE, CBP, and FPS—Firearms, Armor, and Credentials Tracking System (FACTS)—had the ability to track ammunition primarily through the order and receipt process, and ammunition usage was tracked, sometimes manually, at the field level before balances were updated in FACTS. According to ICE officials, ICE and CBP are working to enhance the ability of the ammunition module within FACTS to develop a more accurate accounting ability that is to allow for precise tracking of ammunition usage at the headquarters level. According to ICE, the enhanced module is planned to be deployed by January 2014. We determined these data were sufficiently reliable to report general trends in usage when rounded to the hundred thousands.

[25] We used data from fiscal years 2011 through 2013, because those years represent the most readily available data for DOJ components within our review—FBI, ATF, and the U.S. Marshals Service. The data from ATF included an estimated number of task force officers per year, so the approximate number of rounds for a DOJ officer could be higher or lower depending on the actual number of task force officers who trained with ATF during those 3 years. A task force officer is a sworn state, county or local law enforcement officer who is formally assigned, on a full-time basis, to an ATF task force. For a full description of our methodology, see appendix I.

[26] The FBI provided inventory data as of January 30, 2013. ATF provided inventory data as of May 1, 2013. The U.S. Marshals Service was unable to provide current on-hand inventory data during the time allotted to respond during this review, citing its decentralized system of recording ammunition data.

[27] In addition to issuing guidance on inventory management related to firearms and ammunition, DHS has issued guidance that components are to follow regarding the procurement of firearms and ammunition, as well as for other supplies and services. In addition to the Federal Acquisition Regulation, the primary regulation for use by all federal executive agencies in their acquisition of supplies and services with appropriated funds, DHS and its components are guided by the Homeland Security Acquisition Regulation and the *Homeland Security Acquisition Manual*, which are the primary DHS policies for the procurement of ammunition and firearms.

[28] Department of Homeland Security Inspector General, *DHS Controls over Firearms,* OIG10-41, (Washington, D.C.: Jan. 25, 2010).

[29] Specifically, the DHS Inspector General recommended that those policies and procedures include requirements for properly securing firearms, timelines for recording acceptances and transfers in the inventory system, requirements for reporting lost firearms, and

inventory procedures that include having an independent third party observe annual inventories. OIG-10-41.

[30] A firearm record shall, at a minimum, contain authorized asset user, organization code, manufacturer, model number, serial number, type of firearm, acquisition date, acquisition cost, date of receipt, and location.

[31] A quarterly verification involves verifying a firearm issued to an officer matches the firearm serial number in the component's records. Components may conduct the verification during quarterly firearms qualifications.

[32] We reviewed the firearm policy manuals or related documentation from CBP, FLETC, FPS, ICE, TSA, and USSS to determine whether the manuals specifically addressed the five firearm inventory management requirements identified in the DHS directive. We also interviewed component firearms management officials to discuss their firearms inventory management practices.

[33] DHS Manual 119-03-001-01, *Personal Property Asset Management Program Manual*, (Washington, D.C.: 2013). Specifically, the manual states components are to identify sensitive personal property assets (e.g., weapons and information technology equipment with memory capability) and establish additional control and accountability methods commensurate with that sensitivity to safeguard against loss, theft, and misuse. Components are also to identify personal property assets containing material that could be dangerous to public health or safety, or hazardous to the environment if disposed of irresponsibly, and establish special control and accountability methods to ensure the material is managed properly. The manual defines ammunition as a dangerous and hazardous asset and as a sensitive asset. Sensitive assets must be formally accounted for in an accountable system of record and are also subject to an annual physical inventory and a 15 percent random sampling of the firearm inventory by an independent third party.

[34] According to DHS, a consumable asset is an item of supply that is consumed in use (such as paint, rations, water, office supplies, cleaning and preserving materials, and fuel) or that loses its separate identity when merged into another entity (e.g., nuts and bolts, repair parts, spares, construction materials, components and assemblies, etc.). Consumables are considered to be expendable when issued and do not require formal accountability after they have been issued from the system of record. In addition, the Office of the Chief Readiness Support Officer noted that ammunition is a consumable asset that is considered to be expended once it is issued and does not require formal accountability after it has been issued from the system of record.

End Notes for Appendix I

[1] See, for example, GAO, *Homeland Security: DHS has Enhanced Procurement Oversight Efforts, but Needs to Update Guidance,* GAO-12-947 (Washington, D.C.: Sept. 10, 2012), and Congressional Research Service, *Recent Department of Homeland Security Ammunition Solicitations, Contracts, and Purchases,* Memorandum prepared for distribution to more than one congressional office (Washington, D.C.: Mar. 7, 2013).

[2] We did not examine other law enforcement organizations because they were outside the scope of our study and the other organizations have their own unique mission needs.

[3] We excluded FLETC from this analysis because FLETC has few firearm-carrying officers for which it purchases ammunition. Ammunition purchased by FLETC is utilized by all

students who train on firearms at its facilities, including DHS and other federal law enforcement personnel, as well as state, local, tribal, and international law enforcement personnel.

[4] Department of Homeland Security Inspector General, *DHS Controls over Firearms,* OIG-10-41 (Washington, D.C.: Jan. 25, 2010).

End Notes for Appendix II

[1] Commonly used firearms and ammunition refer to the firearms issued and largest amounts of ammunition purchased by the components in our review.

In: Department of Homeland Security … ISBN: 978-1-63321-162-9
Editor: Lewis Calvin

Chapter 2

RECENT DEPARTMENT OF HOMELAND SECURITY AMMUNITION SOLICITATIONS, CONTRACTS, AND PURCHASES*

Congressional Research Service

This memorandum was prepared to enable distribution to more than one congressional office.

This memorandum addresses recent ammunition solicitations issued and contracts awarded by components of the Department of Homeland Security (DHS).

This memorandum provides information to clarify the size and scope of select acquisitions that have garnered public attention, and provides information to place those solicitations and contracts in context. This memorandum is not intended as an exhaustive survey of non-military acquisition of ammunition by the federal government, nor is it an authoritative assessment of the ammunition requirements of any particular component of the federal government.

---

* This is an edited, reformatted and augmented version of a Congressional Research Service memorandum, M-030713, dated March 7, 2013.

NON-MILITARY USE OF AMMUNITION

Many federal government components have law enforcement officers and investigative agents as part of their workforce. Others do not have traditional law enforcement functions but nonetheless have criminal investigators as part of their Inspectors' General functions. Law enforcement officers and investigative agents are authorized to carry firearms, and are required to train in their use and pass certain firearms qualification standards. This training and qualification process in many cases consumes ammunition far in excess of duty consumption.

For example, one local law enforcement agency in the national capital region has its officers qualify twice a year with a pistol, twice a year with a shotgun, and four times on a rifle, if they are issued a long gun. This qualification alone consumes slightly more than 200 pistol rounds per officer annually (if they do not carry an approved backup handgun, which would require additional qualification). Typical law enforcement handgun practice sessions consume 200-300 rounds or more. [1] Specialty units—such as those employing special weapons and tactics, those that work with dogs, and counternarcotics units— train more frequently with their firearms, and often must qualify on a wider range of weapons than their counterparts. Some officers in civilian law enforcement specialty units may shoot 1,000 rounds in training in a week.[2]

DEPARTMENT OF HOMELAND SECURITY (DHS)

DHS inherited the existing structures and ongoing requirements of its individual components when it was established in 2003. It did not begin operations in an environment that allowed for the establishment of department-wide contracting and procurement systems. Starting about five years ago, DHS began to look for ways to consolidate its ammunition procurements—with congressional encouragement. These changes and relatively short timeline mean there are not any data with which to make consistent comparisons of department-wide ammunition solicitations and contracts over DHS's ten-year existence. Nevertheless, the limited data available are discussed below.

Who DHS Buys Ammunition for

According to the 2008 Census of Federal Law Enforcement Officers, DHS employed more than 45% of all federal law enforcement officers with arrest and firearms authority.[3] This report, developed by the Department of Justice (DoJ), indicates that from 2004 to 2008, the number of full-time officers in U.S. Customs and Border Protection rose 33%. Over that same time period, the number of U.S. Immigration and Customs Enforcement officers rose 20%, and Secret Service officers rose more than 9%.[4] All of these types of officers are trained and certified in the use of firearms.

Several other DHS components have ammunition requirements. These include the Coast Guard (USCG), the Federal Air Marshal Service, the Federal Protective Service (FPS), and the DHS Office of the Inspector General (OIG). *Table 1* provides a breakdown of the number of law enforcement officers in most DHS components based on data from the Department of Justice as of September 2008. Data from the Office of Personnel Management (OPM) and DHS confirm that growth in these components continued into 2012.[5]

Table 1. Select DHS Components' Law Enforcement Officers with Firearms Authority (As of September 2008)

Agency	Number of full-time officers
U.S. Customs and Border Protection	37,482
U.S. Immigration and Customs Enforcement	12,679
Federal Protective Service	[900]
U.S. Secret Service (USSS)	5,226
Office of the Inspector General	157
Federal Emergency Management Agency	84
Federal Air Marshal Service	Classified

Source: Bureau of Justice Statistics, Office of Justice Programs, U.S. Department of Justice, Federal Law Enforcement Officers, 2008, June, 2012.

Notes: The Bureau of Justice Statistics analysis was based on the 2008 Census of Federal Law Enforcement Officers, the most recently available data from the Bureau of Justice Statistics. The Federal Protective Service (FPS) was part of U.S. Immigration and Customs Enforcement at the time the underlying survey was made—it is now part of the National Protection and Programs Directorate (NPPD) of DHS. The survey did not include FPS contract guards. Coast Guard personnel, both military and civilian, were not surveyed, and therefore are not included in Table 1.

DHS does not track the total number of law enforcement officers in its ranks, citing a lack of formal definition of the term *law enforcement officer*. However, it does track by job classification the number of personnel required to carry a gun and be trained and certified in its use. *Table 2* shows DHS's outline for most of these job descriptions. These numbers include personnel from U.S. Customs and Border Protection (CBP), U.S. Immigration and Customs Enforcement (ICE), the Secret Service (USSS), civilian employees of the Coast Guard (USCG), the Office of the Inspector General (OIG), and the Federal Emergency Management Agency (FEMA). Not included in *Table 2* are the Transportation Security Administration's (TSA's) Federal Air Marshals, as the number of marshals is classified; and the more than 41,000 uniformed members of the Coast Guard, as they typically purchase their ammunition through Department of Defense (DoD) contracts.

Table 2. DHS Personnel Trained and Certified in Firearms Use
Number of employees, as of February 9, 2013

Job Classification Series	Description	Employee Count
0090 (FPS only)	Security Administration	829
0083	Police	1,482
1801 (ICE only)	General Inspection, Investigation, Enforcement, and Compliance Series	6,645
1811	Criminal Investigating	10,772
1895	Customs and Border Protection	21,688
1896	Border Patrol Enforcement	21,202
Grand Total		62,618

Source: DHS Legislative Affairs e-mail to CRS, February 22, 2013.

The combined training and certification of these personnel, and supply of ammunition for duty use does not constitute the full scope of DHS demand for ammunition. DHS also includes the Federal Law Enforcement Training Center (FLETC), which annually trains roughly 60,000 law enforcement officers and agents from the federal, state and local levels. It has an entire division entirely dedicated to firearms training. The components of DHS and the many other law enforcement organizations that send officers to train at FLETC use different types of ammunition. Therefore, FLETC not only needs to buy ammunition, but it maintains an inventory of a range of round types for use in training.

DHS Purchases of Ammunition

Table 3 shows the total number of rounds of firearms ammunition purchased by DHS, by component over FY2010-FY2012. Unlike contracts or contract solicitations these numbers represent actual purchases.

Table 3. DHS Ammunition Procurement, FY2010-FY2012

	FY2010		FY2011		FY2012	
Component	Rounds Purchased	Cost	Total Rounds Purchased	Cost	Total Rounds Purchased	Cost
CBP	66,248,000	$17,376,248	48,461,000	$13,895,900	36,475,000	$12,255,040
FLETC	17,056,000	$5,507,057	22,735,654	$7,062,254	7,678,946	$2,485,045
FPS/NPPD	2,946,000	$976,621	1,498,000	$517,412	2,804,000	$928,345
ICE	25,212,000	$7,703,308	13,221,000	$4,505,731	28,061,000	$6,633,460
TSA	15,383,000	$2,818,760	7,124,000	$2,010,642	14,864,000	$4,484,581
USCG	17,472,325	$11,465,150	6,782,540	$8,442,495	9,791,274	$8,309,944
USSS	3,997,500	$2,097,971	8,841,860	$1,802,871	3,503,980	$1,439,495
Total DHS	148,314,825	$47,945,115	108,664,054	$38,237,305	103,178,200	$36,535,910

Source: DHS Legislative Affairs e-mail to CRS, February 22, 2013.

The data provided by DHS show declining budgets for ammunition procurement from FY2010 through FY 2012, and an overall decline in the number of rounds purchased.

The downward trend is not universal. CBP had the largest annual expenditures for ammunition reflected in the above table, and was the only component to see its budget for ammunition purchases and actual number of rounds purchased decline each year. Purchase rates for other components show significant variation, falling only to rise again, or vice versa. Ammunition purchases were not always conducted on the basis of anticipated demand in the upcoming budget year alone, but instead combined multiple years' needs in a single purchase. Changes in the average cost per round from year to year are the likely the result of different combinations of rounds being bought each year.

Has DHS Bought Comparatively More Ammunition in Recent Years?

Some have asked whether DHS is purchasing more ammunition now than it has in the past. While over the past three years, the number of rounds purchased has declined, the number of DHS personnel who carry guns and must train and certify in their use overall has grown over the longer time period beginning from the establishment of the department.

The best documented example of this growth is in the size of the Border Patrol, shown in *Table 4*:

Table 4. Number of Border Patrol Agents, FY2004-FY2012

Fiscal Year	Number of Border Patrol Agents	
FY 2004		10,819
FY 2005		11,264
FY 2006		12,349
FY 2007		14,925
FY 2008		17,499
FY 2009		20,119
FY 2010		20,558
FY 2011		20,736
FY 2012		21,370
Growth, FY2004-FY2012	10,551	(97.5%)

Source: DHS Legislative Affairs e-mail to CRS, February 22, 2013; H.Rept. 112-492.

The Border Patrol is a part of CBP, and CBP is the largest non-military purchaser of ammunition in the DHS budget, representing more than a third of the demand in recent years. Therefore, although we do not have data to illuminate levels of ammunition purchase from the first years of DHS, it follows that, holding all other factors equal except for the growth in personnel, DHS would over time need to purchase more ammunition than it has in the past for training, qualification, and field use.

DHS FY2013 Proposed Ammunition Purchases

Table 5 reflects the DHS budget request for ammunition procurement for FY2013. These figures were based on the administration's FY2013 budget submission made in February 2012. If funded, this request would represent an increase of over $700,000 above the level spent in FY2012, but would be almost $1,000,000 below the FY2011 spending level.

Table 5. DHS Ammunition Acquisition: FY2010-FY2012 Expenditures Compared to the FY2013 Budget Request

DHS Components	FY2010 Acquisitions (Actual)	FY2011 Acquistions (Actual)	FY2012 Acquistions (Actual)	FY 2013 Budget Request
CBP	$17,376,248	$13,895,900	$12,255,040	$12,528,146
FLETC	$5,507,057	$7,062,254	$2,485,045	$5,900,000
FPS/NPPD	$976,621	$517,412	$928,345	$470,000
ICE	$7,703,308	$4,505,731	$6,633,460	$5,200,000
TSA	$2,818,760	$2,010,642	$4,484,581	$4,515,552
USCG	$11,465,150	$8,442,495	$8,309,944	$7,400,000
USSS	$2,097,971	$1,802,871	$1,439,495	$1,250,000
Total DHS	$47,945,115	$38,237,305	$36,535,910	$37,263,698

Source: DHS Legislative Affairs e-mail to CRS, February 22, 2013.

Given the late resolution of the FY2013 appropriation and impact of the sequester, the amount DHS components might have available to spend on ammunition in FY2013 cannot be determined at this time. However, it would be highly unlikely that the Department would have the requested level of resources available in FY2013 to purchase ammunition.

DHS Ammunition Stocks

An assertion in media reports is that DHS has billions of rounds of ammunition stockpiled.[6] CRS does not keep a running track of the department's ammunition stocks. However, according to DHS, in November of 2012, departmental components reported 263.7 million rounds of all types of ammunition on hand— slightly more than was purchased in FY2010 and FY2011 combined.[7] The media assertion that DHS has more than 2 billion round of ammunition stockpiled is therefore high by nearly an order of mangitude.

To put that number in further context, CRS analysis of publicly available DHS contracting information shows that in FY2012, over half of DHS procurement actions for firearms ammunition[8] took place in the fourth quarter of the fiscal year. The funds obligated represented 63% of DHS expenditures on that type of ammunition. While examination of dates for completion of these procurements indicate that it is unlikely that all rounds purchased in that time frame would have been delivered by the date of the inventory, it is possible that this number represents a higher than normal total of rounds on hand.

Commonly Mentioned Types of Ammunition

Hollow-Point Ammunition

A significant concern raised about government purchases of hollow-point ammunition is its effect on the human body. Hollow-point rounds expand on impact, thus causing more damage to the human body than other types of rounds. The expansion of the bullet increases the likelihood that the target will absorb more of the bullet's energy, and therefore stop the round, causing more damage to the target, while reducing the chance that the bullet will exit the target and inadvertently harm bystanders.

The propensity for expanding rounds like the hollow-point to cause greater damage to the target prompted the parties to the Hague Peace Conference of 1899 to agree "to abstain from using bullets that expand or flatten easily in the human body" in warfare.[9] This declaration has not led to general international bans on the manufacture, ownership, or use of these types of rounds, however, by governments within their borders. There is no U.S. restriction on the use of hollow-point ammunition by law enforcement agencies.

In fact, hollow-point ammunition is a standard type of round for handguns carried by law enforcement officers. Media reports quote two experts on the widespread use of hollow-point ammunition:

"For practice ammunition, they do not have to be hollow-points, but hollow-points are the normal police round used for duty ammunition due to their ability to stop when they hit an object as opposed to going through it and striking more objects," said William J. Muldoon, president of the International Association of Directors of Law Enforcement Standards and Training.

... John W. Worden, director of the University of Missouri's Law Enforcement Training Institute [said,] "Hollow points are carried by law enforcement all over the country and are probably the preferred type of ammo no matter what caliber."[10]

"High-Powered" Ammunition

CRS research was unable to identify a broadly agreed upon technical definition of "high-powered rifle" or "high-powered ammunition." Generally, the label "high-powered" is applied to a rifle, not a handgun or a shotgun. One definition of a "high-powered rifle" says the term is used to describe "any modern rifle that fires a bullet at high velocity from a hollow cylindrical metal shell by means of igniting a primer and detonating a smokeless powder charge contained within."[11] Others note the rifle must be "full length," others state that the cartridge must be "centerfire"—which is triggered from the firing pin striking a primer in a separate pocket in the center of the cartridge's "head"— rather than "rimfire"—which is triggered when the firing pin strikes the cartridge rim, compressing a primer located there.[12] Modern shooting sports have types of competition known as "high power rifle" and "high power sporting rifle," but the rules for these competitions do not define these terms inasmuch as they note the equipment allowed for the competition in question, which includes an array of military and civilian long arms.[13]

Selected Recent DHS Solicitations and Contracts for Ammunition

The most common misinterpretation in media reports surrounding this issue confuses a solicitation for proposals—an early stage of the procurement process—with the actual purchase. For ammunition, recent solicitations have frequently been for multiple year contracts, on which multiple future purchases will be made, depending on the department's demand for the ammunition in question and the availability of appropriations.

Immigration and Customs Enforcement Contract for .40-caliber Hollow-Point Ammunition

In March 2012, DHS' Immigration and Customs Enforcement (ICE) awarded a five-year contract for a DHS-wide procurement of up to 90 million rounds of .40-caliber hollow-point ammunition per year—a total of 450 million rounds over the life of the contract if every option period is fully exercised.

DHS released the following statement regarding this ammunition contract in 2012:

> A 5-year department-wide contract was awarded to Federal Cartridge, a wholly-owned subsidiary of ATK. The contract allows the department-wide purchase of up to 450 million rounds of this ammunition, which is the quantity projected to be used over the next five years. The contract is one of a number of department-wide contracts for duty ammunition, but it is the primary vehicle for .40 cal. S&W 180 grain ammo. This contract is intended to be used by all components, except USCG, who uses DoD ammunition contracts. This contract is part of the Department's strategic sourcing efforts to combine multiple previous contracts in order to leverage the purchasing power of the entire Department to efficiently procure equipment and supplies.[14]

DHS provided further detail on the previous contracts in February 2013, noting that there had been three contracts for procurement of .40-caliber hollow point ammunition issued in 2008 and 2009: one five-year contract for a total of 375 million rounds of 180-grain ammunition; another five-year contract for a total of 75 million rounds of 155-grain ammunition; and a third five-year contract for a total of 16 million rounds of 135-grain ammunition. These contracts would have, if all option periods were fully exercised, allowed DHS to purchase 466 million rounds of ammunition. However, DHS found very few purchases were made of the lighter-weight ammunition, as officers preferred to train with the weight of ammunition used in the field and in qualifications.[15]

FLETC Solicitation for "750 Million Rounds of High-Powered" Training Ammunition[16]

In July 2012, the Federal Law Enforcement Training Center (FLETC) requested proposals from vendors for training ammunition. Misinformation has circulated describing this contract being for a purchase of 750 million rounds of "high-powered" or hollow-point ammunition.[17] As described below, CRS analysis of the solicitation (HSFLGL-12-B-00003), finds this description to be incorrect on several counts.

According to the breakdown of ammunition types requested over the term of the solicitation, if each term of this solicitation was fully exercised for every ammunition type, FLETC would purchase slightly more than 10,354,000 rounds of ammunition in the first year of the contract, rising to slightly more than 15,158,000 rounds in the final option period of the contract, for a total of 63,258,000 rounds of ammunition over the course of five years.[18] In addition, an amendment was filed to the solicitation in August 2012, clarifying that language on page 25 of the 91-page solicitation should read "The maximum limit for the resultant award(s) is 70 million rounds for life of the five-year contract(s)."[19] As to the categorization of the type of ammunition sought through this solicitation as hollow-point or "high-powered," FLETC's solicitation identified 31 different types of pistol, rifle, and shotgun ammunition. Fourteen of these ammunition types were hollow-point. As noted above, pistol and shotgun ammunition types do not fall under colloquial definitions of "high-powered ammunition."

FLETC Contract for "21.6 Million Rounds" of Training Ammunition"[20]

In February 2013, FLETC requested proposals for training ammunition. Media accounts cited the number of rounds as 21.6 million.[21] An examination of the solicitation documents, available at the Federal Business Opportunities website (http://FBO.gov) shows several changes to the solicitation after its release, including correction of the classification of ammunition being sought, and clarification of language in the summary description.

The description of what was being sought included the total number of rounds, followed by a statement that the price is to be provided for 1,000 round lots. The next column showed the number of 1,000 round lots to be purchased. The unknown author[22] of the 21.6 million round assertion seems to have multiplied the total sought from the first column by the number of 1,000 round lots to be purchased. Hence two requests for 100,000 rounds become 10 million each, and a request for 40,000 rounds becomes 1.6 million (40,000 x 40), for a mistaken total of 21.6 million.

The amended solicitation dropped one request for 100,000 rounds, and the amended summaries made it clear that the solicitation clearly sought 140,000 rounds. The contract was awarded on February 21, 2013, for less than $40,000.

End Notes

[1] E-mail to analyst from local law enforcement official, February 25, 2013.

[2] Ibid.

[3] The total number of DHS law enforcement personnel is not available, as the number of Federal Air Marshals is classified, and the Census did not include Coast Guard personnel with law enforcement duties.

[4] Bureau of Justice Statistics, Office of Justice Programs, U.S. Department of Justice, Federal Law Enforcement Officers, 2008, June, 2012, p.2

[5] This is evident from FedScope data on employment levels and classifications. However, the FedScope data are a tally of employed individuals, not the number of full-time equivalents (FTEs). Also, FedScope does not directly indicate which positions are law enforcement officers and/or carry firearms. Therefore meaningful direct comparisons of numbers between FedScope data and the Justice Department data are not possible.

[6] See, for example, "DHS adds 21.6 million rounds of ammo to its nearly 2 billion round stockpile," February 7, 2013. http://www.examiner.com/article/dhs-adds-21-6-million-rounds-of-ammo-to-its-nearly-2-billion-round-stockpile.

[7] E-mail to CRS from DHS Legislative Affairs, February 22, 2013.

[8] According to data from the Federal Procurement Data System, this product code (1305) covered almost 80% of the value of DHS ammunition contracts in FY2012.

[9] Declaration III of The Hague Convention of 1899, as found at http://avalon.law.yale.edu/19th_century/dec99-03.asp.

[10] Stephen Ohlemacher, "Social Security Shoots down Web Theories on Ammo," The Washington Post, September 5, 2012, p. B4.

[11] www.firearmstruth.com/2010/media-misinformation-high-powered.

[12] www.shootingusa.com/GUN_QUESTIONS/gun_questions.html.#25.

[13] See rule books for "High Power Rifle" and "High Power Sporting Rifle" at http://compete.nra.org/official-nra-rule-books.aspx.

[14] E-mail to CRS from ICE Legislative Affairs, April 25, 2012.

[15] E-mail to CRS from DHS Legislative Affairs, February 22, 2013.

[16] Information in this section is an analysis of Solicitation Number HSFLGL-12-B-00003, as found at https://www.fbo.gov/index?s=opportunity&mode=form&id=7f6a48adb7fe 40b74 89fbbf5b86eaa0e&tab=core&_cview=1.

[17] See, for example, "Homeland Security finishing acquisition of millions of rounds of high-powered ammo", August 13, 2012. http://rt.com/usa/dhs-ammo-rounds-security-560/

[18] Solicitation Number HSFLGL-12-B-0003, solicitation, July 19, 2012, p. 6-24.

[19] Solicitation Number HSFLGL-12-B-0003, amendment 2, August 16, 2012, p. 1-3.

[20] Content in this section is an analysis of Solicitation Number: HSFLAR-13-Q-00014, as found at https://www.fbo.gov/index?s=opportunity&mode=form&tab=core&id=82540780e44 5be72bef7699c0cd872f7&_cview=1.

[21] See, for example, "DHS adds 21.6 million rounds of ammo to its nearly 2 billion round stockpile," February 7, 2013. http://www.examiner.com/article/dhs-adds-21-6-million-rounds-of-ammo-to-its-nearly-2-billion-round-stockpile.

[22] CRS has been unable to identify the original source of the claim made in multiple media reports circulating on the Internet that the solicitation was for 21.6 million rounds.

In: Department of Homeland Security … ISBN: 978-1-63321-162-9
Editor: Lewis Calvin

Chapter 3

DHS Controls Over Firearms*

Department of Homeland Security, Office of Inspector General

Preface

The Department of Homeland Security Office of Inspector General (OIG) was established by the *Homeland Security Act of 2002* (Public Law 107-296) by amendment to the *Inspector General Act of 1978*. This is one of a series of audit, inspection, and special reports prepared as part of our oversight responsibilities to promote economy, efficiency, and effectiveness within the department.

This report addresses the efficacy of the Department of Homeland Security's management and oversight of its components to ensure that personnel are sufficiently safeguarding and controlling firearms. It is based on interviews with employees and officials of relevant agencies and institutions, direct observations, and a review of applicable documents.

The recommendations herein have been developed to the best knowledge available to our office, and have been discussed in draft with those responsible for implementation.

---

* This is an edited, reformatted and augmented version of a report, OIG-10-41, dated January 2010.

We trust this report will result in more effective, efficient, and economical operations. We express our appreciation to all of those who contributed to the preparation of this report.

Richard L. Skinner
Inspector General

ACRONYMS AND ABBREVIATIONS

CBP	Customs and Border Protection
DHS	Department of Homeland Security
FACTS	Firearms, Armor, and Credentials Tracking System
FMFIA	Federal Managers' Financial Integrity Act
GAO	Government Accountability Office
ICE	Immigration and Customs Enforcement
OCAO	Office of the Chief Administrative Officer
TSA	Transportation Security Administration
USCG	United States Coast Guard
USSS	United States Secret Service

EXECUTIVE SUMMARY

The Department of Homeland Security's mission includes law enforcement elements, which require the department, through its components, to manage various types of firearms. Our audit objective was to determine the efficacy of the Department of Homeland Security's management and oversight of its components to ensure that personnel were sufficiently safeguarding and controlling firearms.

The Department of Homeland Security, through its components, did not adequately safeguard and control its firearms. Components reported 289 firearms as lost during FYs 2006 through 2008. Although some reported losses were beyond the officers' control, most losses occurred because officers did not properly secure firearms. The department did not have a specific firearm policy and instead relied on the components to establish specific policies and procedures for managing, safeguarding, and controlling firearms. While some component policies were sufficient, personnel did not always follow them and the department did not require that independent third parties perform firearm inventories. Field offices did not always promptly report lost firearms to component headquarters or keep inventory records updated. Lost firearms pose serious risks to the public and law enforcement officers.

Based on the results of our audit we are making two recommendations to the department to improve controls over firearms. The Department of Homeland Security's Management Directorate and Office of the Chief Administrative Officer concurred with the recommendations. The components we reviewed in detail are already taking actions to correct the issues we identified.

BACKGROUND

Between 2003 and 2007, the Government Accountability Office (GAO) and the Department of Justice Office of Inspector General assessed controls over firearms at 18 federal law enforcement agencies. The reports showed that these agencies faced similar challenges in safeguarding and controlling firearms and suffered losses of firearms under similar circumstances as the Department of Homeland Security (DHS). In some cases, DHS losses were fewer than other agencies, indicating that DHS may have stronger controls over its firearms.

To support its law enforcement mission DHS had over 188,548 firearms in its inventory as of July 2009. The components with firearms include:

- Customs and Border Protection (CBP);
- Federal Emergency Management Administration (FEMA);
- Federal Law Enforcement Training Center (FLETC);
- Immigration and Customs Enforcement (ICE);
- Science and Technology Directorate[1] (S&T);
- Transportation Security Administration (TSA);
- United States Coast Guard (USCG); and
- United States Secret Service (USSS).

Table 1 illustrates component inventories and table 2 illustrates the types of firearms carried by DHS law enforcement officers.

Table 1. DHS Component Inventories as of July 2009[2]

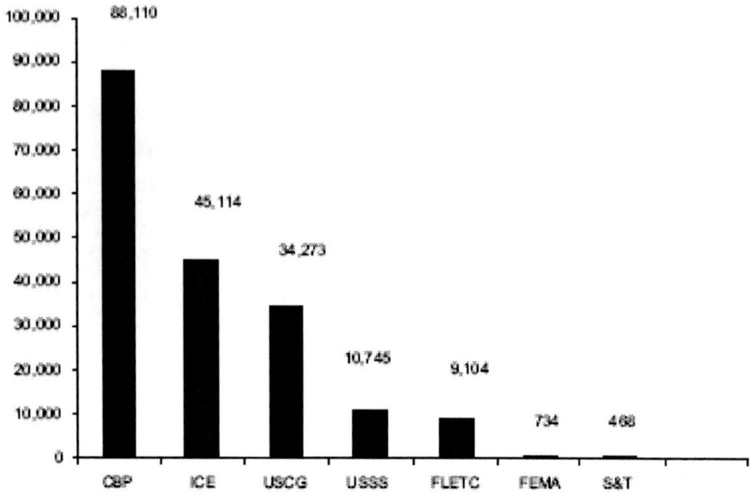

Table 2. Types of Weapons in Component Inventories

Handguns

Image courtesy of CBP

Semiautomatic / Automatic Weapons (M-4 Rifles)

Image courtesy of Colt Defense LLC

Shotguns

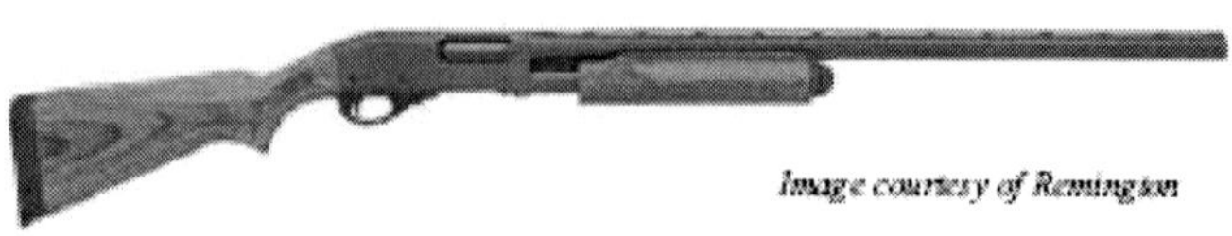

Image courtesy of Remington

The Use of Force Policy Division at CBP and the National Firearms and Tactical Training Unit at ICE oversee their respective component firearms programs. CBP and ICE used the Firearms, Armor, and Credentials Tracking System (FACTS) to manage firearm inventories. CBP transferred its inventory data from the Firearms Inventory Tracking System to FACTS in April 2009. FACTS maintains information relating to individual firearms, such as the serial number, location, date of last inventory, assigned officer, and officer qualification scores.

In 1982, Congress established the *Federal Managers' Financial Integrity Act* (FMFIA) requiring executive agencies to establish and maintain controls that provide reasonable assurance that federal entities safeguard assets against waste, loss, unauthorized use, and misappropriation. The act also mandates that GAO's internal control standards serve as the framework for agencies to use in establishing and maintaining internal control systems. GAO internal control standards provide that in establishing internal controls, agencies should assess the risks associated with asset losses and establish control activities to help ensure those risks are addressed.

In addition to the FMFIA and GAO internal control standards, two other organizations issued advisory criteria addressing inventory controls. In June 1995, the Joint Financial Management Improvement Program addressed management's responsibility to provide guidelines for developing, documenting, and implementing physical controls to safeguard and provide accountability for inventory items. In August 1983, the Commission on Accreditation for Law Enforcement Agencies, Inc., an independent accrediting authority for law enforcement agencies, published accreditation standards that include procedures for inventory and property control. These inventory control requirements apply to sensitive property, such as firearms, to limit accessibility to authorized individuals and ensure accountability.

DHS Management Directive 0565, DHS Personal Property Management, provides a general description of controls for managing property and does not include controls over sensitive items, such as firearms.

We assessed the efficacy of DHS' management and oversight of its components to ensure that personnel were sufficiently safeguarding and controlling firearms. During our audit, we tested the accuracy of inventory records for 1,528 firearms, observed firearm storage procedures, and reviewed inventory policies and procedures at 6 CBP and 10 ICE field offices.

Results of Audit

The department's management and oversight of component safeguards and controls over firearms were not effective. A key reason that firearm controls were not sufficient was because DHS did not have specific firearms policies and procedures in place. Instead, DHS relied on its components to augment its general property management policies and procedures with specific guidance for safeguarding and controlling firearms. Although some component policies and procedures for safeguarding firearms were sufficient, personnel did not always follow them. Component personnel did not always sufficiently safeguard their firearms and, as a result, lost[3] a significant number of firearms between FY 2006 and FY 2008. Most of the losses occurred because law enforcement officers did not sufficiently secure firearms in their possession. The lost firearms created unnecessary risk to the public and law enforcement personnel; in some cases state and local law enforcement officials recovered lost DHS firearms from felons and gang members.

In addition, the department did not require independent third parties to perform annual firearm inventories, component field offices did not always promptly report lost and stolen firearms to component headquarters, and field offices did not keep inventory records updated. As a result, DHS could not ensure that components safeguarded and controlled firearms, or provided accountability over firearms.

During our audit, CBP and ICE initiated changes to their management and oversight of firearms to correct deficiencies we identified during this audit. Both components are reviewing or strengthening policy and procedures for safeguarding firearms and reporting lost firearms.

Safeguarding Firearms

The department, through its components, did not adequately safeguard and control firearms. During FYs 2006 through 2008, DHS components reported 289 handguns, M-4 rifles, and shotguns as lost. CBP and ICE reported 243 (84%) of the 289 lost firearms. The remaining 46 lost firearms (16%) were reported by USCG (9), TSA (22), and USSS (15)[4].

When a law enforcement officer determines that a firearm is missing, CBP and ICE policies require that the officer send a memorandum reporting the loss, and a description of the situation surrounding that loss, to the supervisor.

The supervisor forwards the notification of loss to the respective component headquarters personnel, who create a case file for the lost firearm. We reviewed 243 case files for firearms reported lost by CBP and ICE. For each case file, we determined the reason for the reported loss and evaluated those circumstances against the respective component's policies and procedures.

Lost Firearms

CBP and ICE reported 243 lost firearms during FYs 2006 through 2008. According to our analysis:

- 36 firearms (15%) were lost due to circumstances beyond the control of the officers. For example, CBP lost firearms when Hurricane Katrina made landfall and ICE lost a firearm during an assault on an officer.
- 28 firearms (about 11%) were lost even though officers stored them in lockboxes or safes.
- 179 firearms (74%) were lost because officers did not properly secure them.

Of the 179 firearms, CBP and ICE reported 59 (33%) firearms as lost and 120 (67%) firearms as stolen. Since the components' guidance did not provide a standard methodology for classifying and reporting lost firearms, officers tended to report a firearm as stolen rather than lost. This was due to a common perception among officers that reporting a stolen firearm was more acceptable than reporting a lost firearm.

Although CBP and ICE reported 120 firearms as stolen, our analysis showed that these firearms were lost (stolen) because officers left the firearms unsecured. All 179 losses may have been prevented had the officers properly safeguarded their firearms. For example, an officer left his firearm in the restroom of a fast food restaurant, and when he returned to retrieve it, it was gone. The case file listed this firearm as stolen; however, we believe that the firearm would not have been stolen had the officer exercised due diligence in safeguarding the firearm.

The following examples further demonstrate the inappropriate practices some officers used to store firearms in vehicles and residences:

- A CBP officer left a firearm unsecured in an idling vehicle in the parking lot of a convenience store. The vehicle and firearm were stolen while the officer was inside the store. A local law enforcement officer later recovered the firearm from a suspected gang member and drug smuggler.
- A CBP officer left a firearm on a toolbox in the bed of a truck, and the firearm fell off when the officer drove home. Law enforcement officials later recovered the firearm from an individual who resisted arrest and assaulted the arresting officer.
- An ICE officer left an M-4 rifle and a shotgun unsecured in a closet in his home; subsequently, both firearms were stolen during a burglary. State and federal law enforcement officers later recovered these firearms from a felon.
- An ICE officer left a firearm on the bumper of a vehicle, which fell off as the officer left his place of employment. A civilian found the firearm and turned it over to the local police.

Other CBP and ICE officers left firearms in places such as a fast food restaurant parking lot, a bowling alley, and a clothing store. Although our review focused on CBP and ICE, other components described similar incidents.

For example, a TSA officer left a firearm in a lunch box on the front seat of an unlocked vehicle; the officer realized the firearm was stolen when he returned to the vehicle 2 days later. Officers may have prevented many of these losses had they exercised reasonable care when storing their firearms.

CBP and ICE officers lost 28 firearms even though they secured the firearms in lockboxes or safes. The components may have prevented these losses through more stringent policies and procedures. CBP policies and procedures for leaving a firearm in an unattended vehicle are inadequate and do not include requirements to secure the firearm to the vehicle. Although ICE policies and procedures for storing firearms in vehicles are more stringent, requiring officers to secure them to the vehicle with an ICE-approved safety-locking device, ICE does not require that its staff permanently secure the locking device to the vehicle. While using a safety-locking device is a good practice, the best way to protect firearms from theft is to store them in lockboxes properly mounted as recommended in the manufacturers' guidelines.

STORING FIREARMS AT FIELD OFFICES

Law enforcement officers at field offices did not always use the locking devices provided by CBP and ICE to store their firearms in vehicles. We observed firearm storage procedures and verified 1,528 firearm serial numbers at 6 CBP and 10 ICE field offices. Officers stored 152 of the 1,528 firearms we examined in vehicles. Of the 152 firearms stored in vehicles, 58 (38%), were unsecured. CBP and ICE provided most officers with lockboxes and a locking cable to secure their firearms. However, in these 58 cases, the officers did not use the locking devices. In addition to the locking devices, some ICE facilities provided officers access to lockers to secure their firearms during working hours, as shown below:

Lockers provided for ICE officers to secure firearms – DHS-OIG Photo.

Even though the officers had securing devices and access to these lockers, some officers chose to leave their firearms unsecured in their vehicles. The following examples and photos illustrate how some officers stored firearms in vehicles[5]:

- An ICE officer stored the keys for his assigned government-owned vehicle next to a windshield wiper blade. We observed the officer retrieve the keys, unlock the trunk, and reveal an unsecured firearm, body armor, and radio equipment. After our verification of the firearm serial number, the officer returned the keys to the windshield wiper blade.

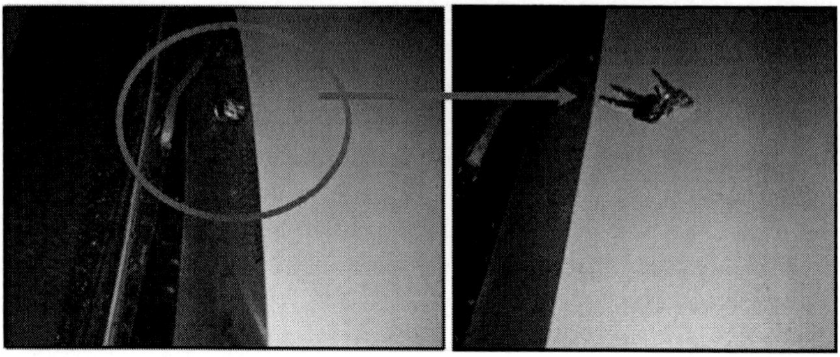

Keys to government- issued vehicle stored next to windshield wiper – DHS-OIG Photo.

- CBP and ICE firearms instructors left firearms unsecured in vehicles. The firearms were stored in bags that were visible from the windows of the vehicles and not secured to the vehicles.
- Other CBP and ICE officers stored firearms in glove compartments and center consoles, under seats, and in various types of bags.

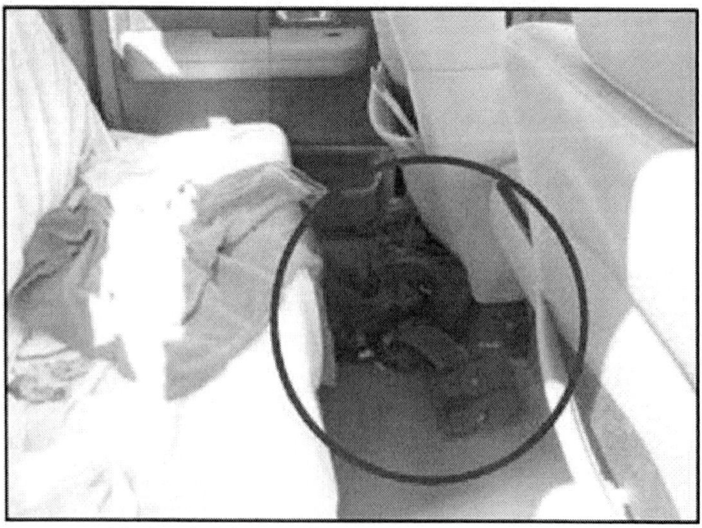

Firearm unsecured in duty belt in an officer's vehicle – DHS-OIG Photo.

Recovered Firearms

According to CBP and ICE case files, 65 (27%) of the 243 lost firearms were recovered; of the 65 recovered firearms, 50 (77%) were recovered by the components or civilian organizations, and 15 (23%) were recovered by other law enforcement organizations. Although lost firearms account for a minor percentage of DHS' total firearms inventory, they pose serious risks to civilians and non-civilians alike. Local law enforcement organizations recovered 15 DHS firearms from felons, gang members, criminals, drug users, and teenagers. For example, law enforcement officers:

- Recovered a firearm from a suspected gang member. The suspect was driving a recreational vehicle with modified hidden compartments that had trace evidence of illegal drugs.

Firearm recovered with gang symbol etched on the barrel.
— Photo courtesy of the Lake City Police Department.

- Recovered a lost firearm from an individual who was in possession of cocaine.
- Recovered a firearm from a drug dealer while executing a narcotics search warrant.

Reporting Lost Firearms

Although officers immediately reported the lost firearms to their supervisors and the National Crime Information Center[6] as required, officers did not always report lost firearms to the component headquarters in a timely manner. In some cases, it took nearly 3 years to report the loss to the component headquarters. According to the case files, CBP and ICE staff took an average of about 2 months and 4 months, respectively, to report lost firearms to component headquarters, ranging from the day of loss to nearly 3 years after the loss. The reporting delays were due to incomplete reporting policies and procedures.

CBP and ICE have incomplete requirements for reporting lost and stolen firearms. Both of their policies and procedures state that when a firearm is lost or stolen the officer must report the loss or theft immediately or as soon as practicable to a supervisor. However, beyond the initial reporting, some critical reporting requirements are not stated. For example, CBP and ICE policies and procedures do not state the timeframes for reporting lost firearms to the National Crime Information Center. (See Appendix C for the table illustrating the incomplete reporting requirements.)

In addition, CBP operates under three inconsistent policies: the former Customs policy, the former Immigration and Naturalization Service policy, and the current ICE policy (only one CBP division follows the current ICE policy). Although ICE has only one policy, it does not address all the critical reporting requirements. Without consistent reporting requirements, the components and department cannot maintain visibility over firearms inventories.

Accounting Inventory Control

Independent third parties do not conduct annual firearms inventories at CBP and ICE. In addition, component personnel did not always enter some acceptances and transfers into the inventory system promptly. CBP and ICE rely on law enforcement personnel and their supervisors to conduct inventories. The delays in acceptances and transfers occurred because CBP and ICE do not have policies and procedures addressing the timelines for these actions. As a result, the inventories were not always accurate or updated.

ANNUAL INVENTORY

Independent third parties do not conduct annual inventories of firearms at CBP and ICE. CBP and ICE guidance does require that field office personnel conduct annual inventories to ensure the completeness and accuracy of the firearms inventory information within the inventory system. The policy requires personnel to complete an annual inventory over a 30-day period. The officer assigned the firearm is responsible for verifying the serial number and other pertinent information, and entering the data into FACTS. Subsequently, the officer's supervisor is responsible for ensuring that the officer appropriately inventoried the correct firearm through visual inspection of the officer's firearm and verification of the information entered into FACTS. We identified six cases of lost firearms where supervisors did not perform a visual inspection; however, they still verified the firearm within the inventory system over several years.[7] The following history of one of these six lost CBP firearms illustrates this annual inventory control weakness:

- September 2004 – An officer's firearm was stolen from his vehicle but the loss was not recorded in the FACTS inventory system.
- August 2005 – Another officer recorded in FACTS that the stolen firearm was in his possession.
- September 2005 – This officer's supervisor affirmed information about this firearm in FACTS.
- December 2006 – A third officer entered the firearm information into FACTS (as if it were not lost).
- December 2006 – This officer's supervisor affirmed the firearm information in FACTS.
- August 2007 – The third officer again entered firearm information into FACTS (as if it were not lost) as part of the annual inventory process.
- September 2007 – Supervisor verified this firearm information in FACTS.

As of 2009, CBP and ICE require supervisors to affirm, within the inventory system, that they visually verified the serial numbers of officers' firearms during the annual inventory. Although this is a reasonable control, affirmations do not replace inventories performed by independent third parties.

ACCEPTANCES AND TRANSFERS

Component personnel did not always promptly update acceptances of firearms and transfers of officers' firearms in the inventory system. Component policies and procedures do not address the timely inputting of acceptances and transfers. Acceptances occur to document the receipt of new firearms. Transfers occur to document a change in an officer's duty station or a change in status of an issued firearm. As a result, the components did not always maintain up-to-date inventory records, which reduced accountability in the management of firearms.

Our review of CBP and ICE resulted in 29 inventory control inaccuracies (22 at CBP and 7 at ICE). For example:

- A CBP facility did not include three firearms issued to the facility in their inventory system. CBP updated the inventory when we pointed out the discrepancy.
- An ICE officer transferred from one office to another but the officer's information, including inventory data for his firearm, did not transfer to the new inventory records for several weeks. ICE updated the inventory records when we pointed out the discrepancy.

ACTIONS TAKEN BY CBP AND ICE

CBP is developing and implementing changes in its management and oversight of firearms to improve firearms accountability and to correct deficiencies identified during our audit. To correct the deficiencies, CBP initiated the following actions:

- Transitioned to a web-based firearms accountability system in April 2009.
- Created a monthly review process of lost firearms to improve its oversight and accountability.
- Developed a unified Use of Force Policy, which is currently in the staffing and approval process.
- Reviewed policies and procedures regarding the safeguarding and controlling of firearms in vehicles to determine if they are sufficient.

- Utilized the CBP*net* to remind law enforcement personnel of the reporting requirements for lost, stolen, or missing firearms.
- Developed a presentation that addresses the proper methods for safeguarding and controlling firearms, which CBP plans to use annually.

ICE initiated changes in its management and oversight of firearms to correct deficiencies identified during our audit. To correct the deficiencies, ICE initiated the following actions:

- Developed methods to promote firearms security awareness through policy and procedure clarifications, and posters to reinforce firearms storage requirements.
- Drafted a memorandum to remind all ICE program office directors of their responsibility to ensure that armed personnel safeguard all issued or approved personally owned firearms in accordance with ICE policies and procedures.
- Drafted a second memorandum to provide supplementary guidance regarding the processes officers should use to report lost and stolen firearms and the required documentation for transfer and receipt of firearms.
- Recommended that its Office of Professional Responsibility Management Inspection Unit expand the scope of their firearms field audits to include a review of the individual officers' storage practices.

CONCLUSION

The prescribed guidance for DHS and the augmented guidance developed by the components are incomplete and inconsistent to ensure that firearms are protected against loss and unauthorized use. By developing standard policies and procedures, DHS could reduce the number of lost firearms and improve its accountability over its firearm inventory.

RECOMMENDATIONS

We recommend that the Undersecretary for Management:

Recommendation #1: Develop department-wide policies and procedures for safeguarding and controlling firearms. At a minimum, these policies and procedures should include:

a) Requirements for properly securing firearms, including a requirement to use properly mounted lockboxes.
b) Timelines for recording acceptances and transfers in the inventory system.
c) Requirements for reporting lost firearms, including classification of lost versus stolen, and timelines for reporting lost firearms to supervisors, local law enforcement, the National Crime Information Center, and component headquarters.
d) Inventory procedures that include having an independent third party observe annual inventories.

Recommendation #2: Assess firearm security equipment needs for each officer assigned a firearm, issue security equipment as needed, and reaffirm to each officer the requirement to always properly secure firearms.

MANAGEMENT COMMENTS AND OIG ANALYSIS

DHS provided comments to our draft report, concurring with both recommendations. Below is a summary of the written response from DHS and our analysis of the response. A copy of the Undersecretary of Management's response, in its entirety, appears in Appendix B. *DHS Response to Recommendation #1*: The Undersecretary of Management concurred with our recommendation and said that the Office of the Chief Administrative Officer (OCAO) has begun a complete revision of the DHS Property Management policy directive. Key controls addressed within the DHS Property Management policy directive will include the proper storage of weapons; recording and updating firearms inventory; classification of lost and stolen, and timeliness for reporting lost or stolen firearms; and proper inventory procedures. *OIG Evaluation*: This recommendation is resolved, but remains open pending confirmation that the DHS implemented these key controls and that the key controls address the recommendation. Developing effective policies and procedures will assist in mitigating the risk of lost firearms.

DHS Response to Recommendation #2: DHS concurred with our recommendation. The OCAO will review the component's requirements for firearm security equipment and collaborate with them to determine the effectiveness of their current equipment. In addition, the OCAO will require that all components conduct annual firearm security requirements awareness training. *OIG Evaluation:* This recommendation is resolved, but remains open pending confirmation that DHS implemented these policies and procedures and that the policies and procedures appropriately address the recommendation. Developing effective policies and procedures will assist in mitigating risks associated with maintaining security over firearms.

APPENDIX A. PURPOSE, SCOPE, AND METHODOLOGY

This report provides the results of our work to determine the efficacy of DHS management and oversight of component safeguards and controls over firearms. To achieve our objectives we:

- Interviewed DHS officials regarding management of personal property, specifically firearms, and the related policies and procedures for DHS and the components;
- Reviewed DHS organizational charts, and manuals for tracking systems and firearms inventories;
- Reviewed case files and data for the components' lost firearms from FYs 2006 through 2008;
- Contacted law enforcement agencies and the Federal Bureau of Investigation to obtain information regarding the circumstances surrounding the recovery of lost firearms;
- Interviewed officers, observed the storage of firearms, and physically verified that serial numbers and the make and model of the officers' firearms matched the inventory listing provided to us by CBP and ICE;
- Conducted fieldwork at 16 CBP and ICE locations, verified the serial numbers for 1,528 issued firearms, and observed storage practices for 152 firearms;
- Reviewed prior audit reports regarding DHS, the components, and other federal agencies that have armed law enforcement officers; and
- Assessed the reliability and validity of data provided by the components by performing a physical verification of inventory data and comparing relevant data from multiple sources.

We conducted our audit between February and July 2009 under the authority of the *Inspector General Act of 1978*, as amended, and according to generally accepted government auditing standards. Those standards require that we plan and perform the audit to obtain sufficient, appropriate evidence to provide a reasonable basis for our findings and conclusions based on our audit objectives. We believe that the evidence obtained provides a reasonable basis for our findings and conclusions based on our audit objectives.

APPENDIX B. MANAGEMENT COMMENTS TO DRAFT REPORT

U.S. Department of Homeland Security
Washington, DC 20528

DEC 11 2009

Homeland Security

MEMORANDUM FOR: Anne L. Richards
Assistant Inspector General for Audits

FROM: Elaine C. Duke
Under Secretary for Management

SUBJECT: Draft Report: DHS Controls Over Firearms

Thank you for the opportunity to respond to the findings and recommendations in the Office of Inspector General (OIG) draft report titled, "DHS Controls Over Firearms" which was transmitted via your memorandum dated October 23, 2009. The report addresses the efficacy of the Department of Homeland Security's management and oversight of its Components to ensure that personnel are sufficiently safeguarding and controlling firearms.

The Management Directorate and the Office of the Chief Administrative Officer (OCAO) acknowledge the importance of effective controls over firearms and other sensitive assets and we are committed to improving current Department-wide policies and procedures, as well as strengthening OCAO oversight of the property management program throughout the Department. OCAO worked with each of the Components that was audited (Customs and Border Protection [CBP], Immigration and Customs Enforcement [ICE], and Transportation Security Administration [TSA]) in order to identify corrective actions and any concerns relative to the draft report. We believe this demonstrates our commitment to the Secretary's One DHS initiative and that it will help improve management oversight of a DHS-wide program.

The draft audit report recommendations and our responses and comments follow:

Recommendation #1: Develop department-wide policies and procedures for safeguarding and controlling firearms. At a minimum, these policies and procedures should include:

> **a. Requirements for properly securing firearms, including a requirement to use properly mounted lockboxes.**

We concur with this recommendation. OCAO has begun a complete revision of the DHS Property Management policy directive. In addition, OCAO will initiate a comprehensive review of current component requirements for securing firearms and other sensitive assets with the goal of identifying best practices to secure, safeguard and control these assets. Several key controls will be addressed in the department-wide policies and procedures including:

- Proper storage of weapons;
- Ensuring accountability for safeguarding and controlling firearms;

- Standard training for Component level personnel who have responsibility for firearms to raise awareness of the accountability aspects of DHS firearms controls;
- Requirements for investigation when determinations are made that individuals do not follow accountability procedures.

b. Timelines for recording acceptances and transfers in the inventory system.

We concur with this recommendation. It should be noted that of all the Components that use and store weapons, only two (CBP and ICE) use the same inventory system to manage their weapons. However, OCAO will initiate DHS-wide inventory controls to ensure written policies and procedures are implemented to require accurate and timely firearms property records and systems are maintained. The policies and procedures will include specific requirements to:

- Record firearms upon receipt;
- Update firearms inventory data upon removal of firearms from agency for losses and disposals;
- Update firearms inventory data to reflect discrepancies after on-hand inventory counts.

c. Requirements for reporting lost firearms, including classification of lost versus stolen, and timelines for reporting lost firearms to supervisors, local law enforcement, the National Crime Information Center, and component headquarters.

We concur with this recommendation. OCAO has initiated a review of the current definitions and use of these terms. A consistent DHS-wide policy regarding the classification of lost and stolen, and the timelines for reporting lost or stolen firearms, will be included in the revised Property Management policy directive and distributed throughout the Department.

d. Inventory procedures that include having an independent third party observe annual inventories.

We concur with this recommendation. Due to the diverse nature of the DHS Component operations, third party participation in the annual inventories may be achieved in a variety of different ways. OCAO will review current component inventory policies and procedures and issue instructions regarding proper inventory procedures.

Recommendation #2: Assess firearm security equipment needs for each officer assigned a firearm, issue security equipment as needed, and reaffirm to each officer the requirement to always properly secure firearms.

We concur with this recommendation. OCAO will review Components' requirements for firearm security equipment and collaborate with them to determine the effectiveness of their current equipment. OCAO will require that all components conduct annual firearm security requirements awareness training for all personnel issued firearms. The training will emphasize the importance of properly securing firearms and the risks associated with failing to maintain effective custody of DHS issued firearms.

As noted in the draft report, the Components that were audited took immediate and substantive actions to correct noted deficiencies and to improve the overall management of their firearms. OCAO has initiated action to revise its policies and procedures, with the expectation that a new policy directive will be issued by the spring of 2010. Concurrent with that effort, OCAO will develop a Management Action plan by December 30, 2009, to identify remaining actions, milestones and resource requirements.

APPENDIX C. REPORTING LOST FIREARMS POLICIES AND PROCEDURES

	CBP Policies		*ICE Policy*
Critical Reporting Steps	*Customs Policy*	*Immigration and Naturalization Service Policy*	*ICE Interim Policy*
Report lost firearm to supervisor	Immediately	As soon as practical	Immediately
Report lost firearm to local law enforcement	Not stated	As soon as possible from supervisor	Not required
Report lost firearm to component headquarters	No timeframe	Within 24 hours	Immediately from supervisor
Report lost firearm to National Crime Information Center	No timeframe	Not stated	No timeframe
Memo prepared by employee	Not stated	Within 48 hours	Within 48 hours
Report to Joint Intake Center (JIC)	Not stated	Not stated	Not stated
Significant Incident Report	Not stated	Within 24 hours	Not stated
Report of survey	No timeframe	No timeframe	Not stated
CBP - Internal Affairs	No timeframe	Within 24 hours by HQs	Not applicable
ICE - Office of Professional Responsibility	Not applicable	Not applicable	As soon as practicable from supervisor

APPENDIX D. MAJOR CONTRIBUTORS TO THIS REPORT

Linda Howard, Director
Sean Pettersen, Audit Manager
David DeHaven, Auditor-in-Charge
Lindsay Cabral, Auditor
Brian Blaha, Auditor
Victoria Phan, Program Analyst
Kevin King, Program Analyst
Joseph Faulk, Program Analyst
Lisa Vonder Haar, Desk Officer

End Notes

[1] The Science and Technology Directorate uses firearms solely in association with the testing mission of the Transportation Security Laboratory and does not issue side-arms to individuals.

[2] We did not include the number of firearms in TSA's inventory because this information is protected as Sensitive Security Information under 49 CFR 1520.5(b)(8)(ii). We also did not list the total number of firearms in DHS' inventory to protect TSA's Sensitive Security Information.

[3] Lost firearms are those no longer in the possession of the components; they include firearms that were lost or stolen.

[4] According to Secret Service officials, 3 of the 15 firearms reported as lost or stolen between 2006 and 2008 were actually lost in the 2001 attack on the World Trade Center. The loss of these firearms was reported in 2008.

[5] We notified local supervisors of these issues as we identified them.

[6] The National Crime Information Center is a computerized database for ready access by law enforcement agencies to assist in apprehending fugitives, locating missing persons, locating and returning stolen property, and protecting the law enforcement officers encountering the individuals described in the system.

[7] We referred these six cases to the CBP Office of Internal Affairs for further review.

INDEX

A

B

C

D

E

F

T

U

V

W